T0332652

Advanced AI Techniques and Applications in Bioinformatics

Smart and Intelligent Computing in Engineering

Series Editor:
Prasenjit Chatterjee, Morteza Yazdani, Dragan Pamucar, and Dilbagh Panchal

Artificial Intelligence Applications in a Pandemic
COVID-19
Salah-ddine Krit, Vrijendra Singh, Mohamed Elhoseny, Yashbir Singh

Advanced AI Techniques and Applications in Bioinformatics
Edited by Loveleen Gaur, Arun Solanki, Samuel Fosso Wamba, Noor Zaman Jhanjhi

For more information about this series, please visit: https://www.routledge.com/our-products/book-series

Advanced AI Techniques and Applications in Bioinformatics

Edited by
Loveleen Gaur
Arun Solanki
Samuel Fosso Wamba
Noor Zaman Jhanjhi

CRC Press
Taylor & Francis Group
Boca Raton London New York

CRC Press is an imprint of the
Taylor & Francis Group, an **informa** business

First edition published 2022
by CRC Press
6000 Broken Sound Parkway NW, Suite 300, Boca Raton, FL 33487-2742

and by CRC Press
2 Park Square, Milton Park, Abingdon, Oxon, OX14 4RN

CRC Press is an imprint of Taylor & Francis Group, LLC

ISBN: 978-0-367-64169-6 (hbk)
ISBN: 978-0-367-64767-4 (pbk)
ISBN: 978-1-003-12616-4 (ebk)

DOI: 10.1201/9781003126164

Typeset in Times
by Deanta Global Publishing Services, Chennai, India

Contents

Contributors

Debabrat Baishya
Department of Bioengineering and
  Technology
Gauhati University
Guwahati, Assam, India

Gaber El-Saber Batiha
Department of Pharmacology and
  Therapeutics
Faculty of Veterinary Medicine
Damanhour University
Damanhour, Egypt

Kaushik Kumar Bharadwaj
Department of Bioengineering and
  Technology
Gauhati University
Guwahati, Assam, India

Anuradha Bhardwaj
Department of Biotechnology
Gautam Buddha University
Greater Noida, Uttar Pradesh, India

Harshit Bhardwaj
Department of Computer Science and
  Engineering
University School of Information and
  Communication Technology
Gautam Buddha University
Greater Noida, Uttar Pradesh, India

Divya Bhatia
Department of Biotechnology
University Institute of Engineering and
  Technology
Kurukshetra University
Thanesar, Haryana, India

David Correa Martins-Jr
Center of Mathematics, Computing, and
  Cognition

Federal University of ABC
Santo André, São Paulo, Brazil

Jayashankar Das
Centre for Genomics and Biomedical
  Informatics
Siksha 'O' Anusandhan Deemed to be
  University
Bhubaneswar, Odisha, India

Jagjit Singh Dhatterwal
Computer Science and Applications
PDM University
Bahadurgarh, Haryana, India

Javaria Fazal
Department of Biochemistry
University of Agriculture Faisalabad
Punjab, Pakistan

Vicente Garcia Diaz
Department of Computer Science
University of Oviedo
Asturias, Spain

Loveleen Gaur
Amity International Business School
Amity University
Noida, Uttar Pradesh, India

Arabinda Ghosh
Microbiology Division
Department of Botany
Gauhati University
Guwahati, Assam, India

NZ Jhanjhi
School of Computer Science and
  Engineering
Taylor's University
Selangor, Malaysia

S. Karthikeyan
Department of Computer Science
Banaras Hindu University
Varanasi, Uttar Pradesh, India

Kuldeep Singh Kaswan
School of Computing Science and
  Engineering
Galgotias University
Greater Noida, Uttar Pradesh, India

Love Kumar
Department of Electronics and
  Communication
DAV Institute of Engineering and
  Technology
Jalandhar, Punjab, India

Saswati Mahapatra
Department of Computer Application
Siksha 'O' Anusandhan Deemed to be
  University
Bhubaneswar, Odisha, India

Deepak Malik
Department of Biotechnology
University Institute of Engineering and
  Technology
Kurukshetra University
Thanesar, Haryana, India

Misbah Manzoor
Centre of Agriculture Biochemistry and
  Biotechnology
University of Agriculture Faisalabad
Punjab, Pakistan

Vikrant Nain
Department of Biotechnology
Gautam Buddha University
Greater Noida, Uttar Pradesh, India

Memood Naqvi
Department of Engineering and
  Technology
Mohawk College
Hamilton, Canada

Abhigyan Nath
Department of Biochemistry
Pt. Jawahar Lal Nehru Memorial
  Medical College
Raipur, Chhattisgarh, India

Pradeep N
Dept. of Computer Science and
  Engineering
Bapuji Institute of Engineering and
  Technology
Davangere, Karnataka, India

Haleema Qamar
Department of Biomedical Engineering
Binghamton University
Binghamton, NY

Bijuli Rabha
Department of Bioengineering and
  Technology
Gauhati University
Guwahati, Assam, India

Aamna Rafique
Department of Biochemistry
University of Agriculture Faisalabad
Punjab, Pakistan

Rehab A. Rayan
Department of Epidemiology
High Institute of Public Health
Alexandria University
Alexandria, Egypt

Roopa GM
Department of Computer Science and
  Engineering
Bapuji Institute of Engineering and
  Technology
Davangere, Karnataka, India

Soobia Saeed
Department of Software Engineering
Universiti Teknologi Malaysia
Johor Bharu, Malaysia

Aditi Sakalle
Department of Computer Science and
    Engineering
University School of Information and
    Communication Technology
Gautam Buddha University
Greater Noida, Uttar Pradesh, India

Anurag Sharma
Faculty of Engineering, Design, and
    Automation
GNA University
Phagwara, Punjab, India

Uttam Sharma
Department of Computer Science and
    Engineering
University School of Information and
    Communication Technology
Gautam Buddha University
Greater Noida, Uttar Pradesh,
    India

Vikrant Sharma
Faculty of Engineering, Design, and
    Automation
GNA University
Phagwara, Punjab, India

Sushant Kumar Shrivastava
Department of Pharmaceutical
    Engineering and Technology
Indian Institute of Technology
Varanasi, Uttar Pradesh, India

Shryavani K
Department of Computer Science and
    Engineering

Bapuji Institute of Engineering and
    Technology
Davangere, Karnataka, India

Dhiraj Sinha
Institut Hospitalo-Universitaire
Aix Marseille University
Marseille, France

Arun Solanki
Department of Biotechnology
Gautam Buddha University
Greater Noida, Uttar Pradesh,
    India

Tripti Swarnkar
Department of Computer Application
Siksha 'O' Anusandhan Deemed to be
    University
Bhubaneswar, Odisha, India

Raghunath Satpathy
School of Biotechnology
Gangadhar Meher University
Sambalpur, Odisha, India

Pradeep Tomar
Department of Computer Science and
    Engineering
University School of Information and
    Communication Technology
Gautam Buddha University
Greater Noida, Uttar Pradesh, India

Manish Kumar Tripathi
Department of Pharmaceutical
    Engineering and Technology
Indian Institute of Technology
Varanasi, Uttar Pradesh, India

Abeedha Tu-Allah Khan
School of Biological Sciences
University of the Punjab
Lahore, Punjab, Pakistan

Qurat Ul Ain
Department of Chemistry
Government College Women
   University
Faisalabad, Pakistan

Imran Zafar
Department of Bioinformatics and
   Computational Biology
Virtual University Pakistan
Punjab, Pakistan

Editors

Prof. Loveleen Gaur is the Professor and Program Director (Artificial Intelligence and Business Intelligence and Data Analytics) of the Amity International Business School, Amity University, Noida, India. She is a senior IEEE member and series editor with CRC Press and Wiley. Prof. Gaur is an established author and researcher; she has filed five patents and two copyrights in AI-IoT. For over 18 years she served in India and abroad in different capacities. She has specialized in the fields of artificial intelligence, the Internet of Things, data analytics, data mining, and business intelligence. Prof. Gaur has pursued research in truly inter-disciplinary areas and has authored and co-authored books with renowned international and national publishers like Elsevier, Springer, and Taylor & Francis. She was invited to be guest editor for reputed Springer and Emerald Q1 journals. She has published many research papers in SCI and Q1 journals. She has chaired various positions in international conferences of repute and is a reviewer with top-rated journals of IEEE, SCI, and ABDC.

Dr. Arun Solanki is an Assistant Professor in the Department of Computer Science and Engineering at Gautam Buddha University, Greater Noida, India. He received his Ph.D. in Computer Science and Engineering from Gautam Buddha University in 2014. He has supervised more than 60 dissertations and has published many research articles in Scopus indexed international journals. Dr. Solanki is the editor of several published books and a reviewer in journals published by reputed publishers.

Dr. Samuel Fosso Wamba is a Professor at Toulouse Business School in France. He earned his Ph.D. in industrial engineering at the Polytechnic School of Montreal, Canada. His current research focuses on SCM, E&M-commerce, IT-enabled social inclusion, blockchain, social media, business analytics, big data, and artificial intelligence. He is academic co-founder of RFID Academia, and founder and CEO of e-m-RFID. He has published papers in IS and OM journals and conferences. He is the guest editor of various special issues on IT/RFID/big data for various IS and OM conferences and journals. He is the head of the newly created Artificial Intelligence and Business Analytics Cluster at Toulouse Business School.

Dr. Noor Zaman Jhanjhi is an Associate Professor at Taylor University, Malaysia, with 20 years of rich experience. He is an associate editor for several reputable journals including *IEEE Access* and PC, and a board member for several IEEE conferences around the globe. He is an active reviewer for a series of quality journals. He has authored several research papers in WoS/ISI indexed journals and in impact factor research IEEE journals, has edited 12 books in computer science, and has supervised a number of postgraduate students. He has successfully completed more than 19 international funded research grants.

1 An Artificial Intelligence-based Expert System for the Initial Screening of COVID-19

Anurag Sharma, Hitesh Marwaha,
Vikrant Sharma and Love Kumar

CONTENTS

1.1 INTRODUCTION

Diverse types of viruses exist; unlike other biological entities, several viruses, like poliovirus, have RNA genomes, herpes virus has DNA genomes, and the influenza virus has a single-stranded genome, while others like smallpox have double-stranded genomes [1]. Similarly, coronaviruses consist of a positive-sense single-stranded RNA genome and a nucleocapsid of helical symmetry. During the occurrence of an extremely infectious virus with human-to-human transmission, hospitals and doctors have increased workloads and inadequate resources to cure and hospitalize suspected patients. COVID-19 is a type of harmful virus which in severe cases results in death. There is no vaccine to date to cure this deadly virus. The only way to

DOI: 10.1201/9781003126164-1

1

protect ourselves from COVID-19 is to quarantine ourselves and maintain social distance. An increased number of patients with coronavirus were diagnosed in Wuhan, Hubei Province, China from December 2019. Apart from China, COVID-19 cases were also reported in other parts of the world and it became a pandemic. WHO was informed of patients with pneumonia of unidentified etiology found in the city of Wuhan in China on 31st December 2019. By 3rd January 2020, a total of 44 patients suffering from pneumonia were reported by WHO. In this report, the causative agent was unidentified [2].

On 11th and 12th January 2020, WHO acknowledged a thorough report from the National Health Commission in China that stated that the outburst was linked with the exposure of seafood in a market in Wuhan. The Chinese authorities recognized it as a coronavirus, which was out of control on 7th January 2020. Further, on 12th January 2020, the government of China reported the genetic sequence of the novel coronavirus for countries to use in making specific diagnostic kits. By 20th January 2020, 286 confirmed cases had been reported from China, Japan, Thailand, and the Republic of Korea. The WHO report of China by 10th April 2020 reached up to 83,305 confirmed cases of COVID-19 [2].

The WHO reported 1,610,909 cases (92,798 deaths) of COVID-19 globally by 10th April 2020. Till now, the highest numbers of cases were found in the United States of America. It has reached up to 425,889 confirmed cases of COVID-19 (14,665 deaths) by 11th April 2020. To date, the number of patients suffering from COVID-19 is increasing at a higher rate. Due to the unavailability of effective vaccines for COVID-19, precautions like personal hygiene and quarantine measures have become most important to break this series of communication and slow down the pace of the outbreak [3].

Therefore, the Indian government declared a complete lockdown for three weeks on 22nd March 2020 to combat the spread of COVID-19 in India. By 10th April 2020, India reached up to 6,412 COVID-19 cases (199 deaths). As COVID-19 is increasing rapidly, the Indian government decided to extend the lockdown period till 30th April 2020. By 24th July 2020, the total cases reported in India were more than 12,000,000 [4].

This work aims to develop an expert system based on fuzzy logic which can be used for initial self-screening of COVID-19. Moreover, the hierarchical fuzzy technique has been used to reduce the complexity of the designed system. The idea is to reduce the burden on hospitals and doctors by developing a self-screening tool for the initial diagnosis of COVID-19.

Figure 1.1 shows the global data of confirmed COVID-19 cases and deaths over time till 18th May 2020. The graph shows the perpetuating data of individuals day by day.

1.2   REVIEW OF LITERATURE

To the best of our knowledge there is not much literature and researchers have not contributed much on COVID-19 to date; some researchers have reviewed and discussed COVID-19, summarized as follows:

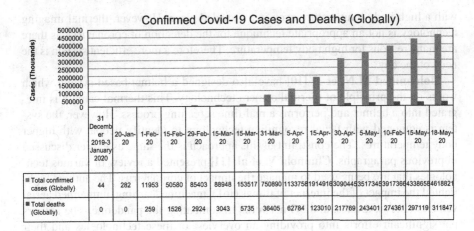

	31 December 2019-3 January 2020	20-Jan-20	1-Feb-20	15-Feb-20	29-Feb-20	15-Mar-20	15-Mar-20	31-Mar-20	5-Apr-20	15-Apr-20	30-Apr-20	5-May-20	10-Feb-20	15-May-20	18-May-20
Total confirmed cases (Globally)	44	282	11953	50580	85403	88948	153517	750890	1133758	1914916	3090445	3517345	3917366	4338658	4618821
Total deaths (Globally)	0	0	259	1526	2924	3043	5735	36405	62784	123010	217769	243401	274361	297119	311847

FIGURE 1.1 WHO 2020 report of confirmed COVID-19 cases and deaths [3].

Chatterjee et al. [5] reviewed the budding proofs to assist and guide the public health reaction, primarily in India. This research has outlined the criteria to be used to produce significant knowledge for recommendations to prevent and control the deadly virus. The paper concluded with the imperative outcomes of continuing hard work to avert and contain COVID-19. Moreover, it points out the requirements for investment in health systems, improved health policy mechanisms, and the requirements for preparations as well as worldwide health security. **Agarwal et al. [6]** described the essentials of the design of such components as space, waste disposal, infection control, and the protection of healthcare personnel, associates concerned in planning and preparation, which can be adjusted to the framework for a new structure or makeshift construction on top of an accessible formation. The researchers finished by giving explicit requirements such as infection avoidance and control to slow the COVID-19 pandemic. **Deng et al. [7]** explained that in a solitary center case sequence of 138 confirmed patients hospitalized in Wuhan, China, the presumed hospital-connected diffusion of coronavirus was suspected in 41% of patients and 26% were in the intensive care unit (ICU), with the deaths of 4.3%. However, gender- and age-based outcomes were also revealed during this study. Further, the authors explained that hospital-linked communication was alleged as the presumed method of infection for health professionals. According to the authors, the median time from contact to first symptom was 5 days and to hospital admission was 7 days. **Mandal et al. [8]** explained that pinpointing quarantine would identify and quarantine 50% of symptomatic persons within 3 to 4 days of increasing symptoms. Moreover, the researchers concluded that the detection of travelers at the access of any harbor with indicative clinical features and from COVID-19-affected countries would achieve a delay in the entry of the virus into the community. **Mohammed M.N. et al [9]** have developed a system for the detection of COVID-19 based on thermal imaging techniques without much human interaction. The authors used internet of things technology to design this system. The purpose of this technology is to detect a human being with a high body temperature even in a crowded place. However, thermal imaging technology is not an appropriate technique for the detection of coronavirus as there are many reasons for high body temperature. Therefore a more efficient system is the need of the day.

Mohammed M.N. et al [10] have also designed a helmet-based system which works on the principle of thermal camera technology. This thermal camera is integrated into a helmet and performs a real-time screening process. However, the system is equipped with facial recognition technology to identify people with higher body temperatures. The constraint of this approach is the same as earlier discussed in previous paragraphs. **Chamola V. et al. [11]** presented a review of various technologies that are being used to manage the impact of coronavirus. The authors highlighted the impact of the internet of things (IoT), machine learning, unmanned aerial vehicles (UAV), drones, and 5G networks in combating the pandemic. The authors put significant efforts into providing an overview of these technologies and their role in the current virus condition. **Waheed A. et al. [12]** presented a new method, namely synthetic chest X-ray (CXR) with the help of *Auxiliary Classifier Generative Adversarial Network* (ACGAN). This method is named CovidGAN. This is proposed to improve CNN for coronavirus detection. The authors claim 95% accuracy with the proposed scheme compared to 85% with CNN. **Basset et al. [13]** proposed an improved marine predators algorithm (IMPA) with an X-ray segmentation-based hybrid COVID-19 detection model. The authors incorporated a ranking-based diversity lessening model to improve the performance of existing IMPA. The authors validated the results with the chest X-ray by assuming the threshold value in the range of 10 to 100 and compared the outcomes with EO, WOA, SCA, HHA, and SSA algorithms. This hybrid approach has given better outcomes compared to other detection methods. Further, **Dong D. et al. [14]** explored the role of AI and Big Data with a quantitative analysis. The review has indicated that the imagining characteristics play an important role in the management and detection of coronavirus. The authors have highlighted the use of quantitative image analysis along with artificial intelligence methods to control, manage, and detect COVID-19. **Quoc-Viet Pham et al. [15]** have emphasized the role of AI and Big data techniques in response to COVID-19. The authors focus on the application area of AI and Big Data to battle against the coronavirus along with the existing confronts and problems linked with the state-of-the-art solutions. The authors also recommended effective control methods to slow the worldwide spread of the COVID-19 pandemic. **Wang N. et al. [16]** presented a COVID-Net by converging deep learning and radiography along with image classifiers. The model is based on transfer learning and model integration. The model is more sensitive than radiography and has achieved 96.1% accuracy and is recommended for use by radiologists to augment the accurateness and competence of the diagnosis of COVID-19.

1.3 MATERIAL AND METHOD

Fuzzy logic is equivalent to a person's feeling and implication process. A fuzzy inference system (FIS) can proficiently model an expert's knowledge for a precise problem in the form of prearranged if-then rules. The ethics of fuzzy modeling were sketched by R.A. Zadeh who introduced the concept of a type of membership and published his study on fuzzy sets that gave birth to fuzzy logic systems. This concept represents one of the most significant paradigms and offers an estimated but valuable means of predicting the outcomes of systems that are too intricate to use mathematical analysis. There is a plethora of applications of fuzzy logic in a large variety of areas like data classification, decision making, pattern recognition, etc. The fuzzy theory gives a practical mode to indicate information linguistically and utilize these linguistic variables for a conclusion mechanism by means of mathematical computations, and these systems are known as *fuzzy rule-based models*. The accomplishment of fuzzy rule-based models is usually an extended action involving several steps like variable selection, knowledge acquisition, membership function selection, the definition of the controller, the definition of rules, etc.

Fuzzy systems exhibit the following characteristics that make them a popular choice for solving different problems:

- These systems can model non-linear behavior exhibited by almost all real systems.
- Fuzzy systems (FS) have proven to be universal approximates, which implies that they are capable of approximating any numeric behavior to the desired accuracy.
- They symbolize facts as combinations of rules that are based on natural language, which is an extremely convenient way to explain decision processes.
- FS articulates ideas with linguistic labels that are close to human intuition.
- They are flexible, and it is easy to add features to them without starting from scratch.
- They provide an alternate design methodology that reduces the design time and produces inexpensive yet robust systems, especially for biomedical applications.

1.3.1 HIERARCHICAL FUZZY SYSTEM

Fuzzy logic systems are quite similar to human beings as they can act according to the sensation and conclusion process which efficiently shapes a specialist's expert opinion for a dedicated hitch in the way of a well-arranged if-then structure. There exist numerous domains for FIS in an enormous variety of fields such as making decisions, recognition of patterns, classification of data, etc.

The fuzzy theory gives various methods for the signification of information and it puts to use these linguistic factors for the inference method by staging numerical computations. These systems are symbolized as *fuzzy rule-based systems*. The system execution is an extended motion including a series of stipulated steps, i.e. acquisition of knowledge, defining of controller and rules, etc. In a classical fuzzy system, the number of rules increases drastically as the number of input variables increases. For example we have input variables "x" and membership function "y"; in that case our fuzzy system is in need of "y^x" rules to develop a *fuzzy rule-based system*. As there is an appraisal in x, the rule base would make the fuzzy rule-based system inevitably difficult to execute. Factually, the dimness of a hitch rises rapidly with input variables, which is referred to as the "curse of dimensionality" by field experts [8].

Therefore, the concept of hierarchical fuzzy systems (HFSs) is introduced to combat the "curse of dimensionality" and the rule outburst issue. These HFSs are comprised of multiple "low-dimensional fuzzy systems" in a hierarchical structure. HFSs are beneficial as with the number of input variables the rules increase only linearly as shown in Figure 1.2.

The purpose of this study was to build up a precise, prompt, and comprehensible screening system to classify the rating of COVID-19, i.e. "Normal," "Proper personal hygiene," and "Seek consultation with a doctor." The major benefit of HFS is to radically diminish the number of rules if input variables have been increased. In HFS, every part is planned liberally and then based on the hierarchical structure; the output of one module is conjoined to the input of subsequent modules.

1.3.2 METHODOLOGY

After doing a literature survey, 15 input parameters were used for the initial screening of COVID-19. These input parameter values are put to use for predicting the status of the body. After the selection of the input variables, the next step is to fuzzify the variables, i.e. the determination of the fuzzy sets for each input parameter and its analogous choice of the belonging to each and every fuzzy rule set. Based on the rules developed in FIS (fuzzy inference system) files gives an accurate result, thus the hitch of MATLAB installation is uninvolved by making the .exe file of our expert system (Figure 1.3).

With the help of screening test parameters used to diagnose COVID-19, 15 input variables were selected for designing the expert system, including cough, cold,

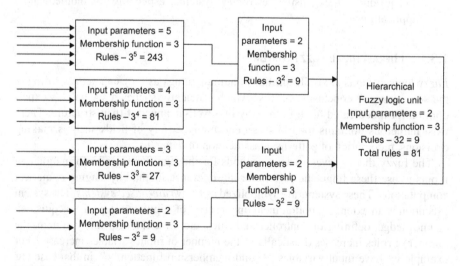

FIGURE 1.2 Hierarchical fuzzy expert system.

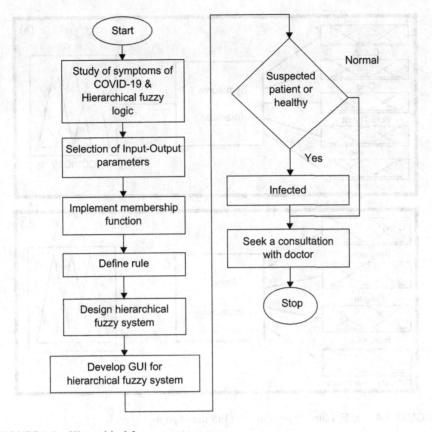

FIGURE 1.3   Hierarchical fuzzy expert system.

sore throat, diarrhea, fever, body aches, fatigue, headache, difficulty in breathing, travel history of the past 14 days, COVID-infected area, direct contact or taking care of COVID patients, age, other medical conditions, and the status of the last 48 hours [9].

1.4   RESULTS

This section describes the results of the designed system based on the various input variables, i.e. cough, cold, fever, etc.

1.4.1   Fuzzy Inference System

Figure 1.4(a) shows the fuzzy inference system (FIS) which includes five input parameters, i.e. cough, cold, sore throat, diarrhea, fever. These are the initial symptoms for COVID-19. Based on these inputs the FIS engine with the help of if-then rules predicts the output.

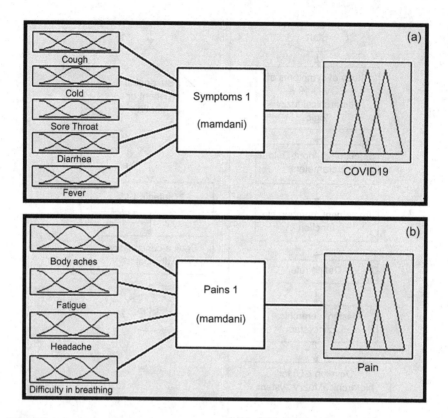

FIGURE 1.4   (a) FIS file of symptoms. (b) FIS file of pains.

Figure 1.4(b) shows the FIS file which further includes four more parameters, i.e. body aches, fatigue, headache, difficulty in breathing. For simplicity, only two FIS files have been demonstrated in this section.

1.4.2   MEMBERSHIP FUNCTIONS

Membership functions [49] are associated with each input variable. As a result, the relationship functions of parameters revealed in this research portray the unambiguous outline of the membership functions. Figure 1.5 shows the membership function of cough which comprises three membership plots, namely: normal or less, moderate, and high. A trapezoidal membership function *trapmf* is operated for the input trait and a trapezoidal membership function is operated for the output attribute. A trapezoidal function is used to decide the full input and output range.

1.4.3   RULE EDITOR

Describing the emergence of the configuration is called a fact list which can be edited by a rule editor. A rule editor encompasses a huge editable textual field for

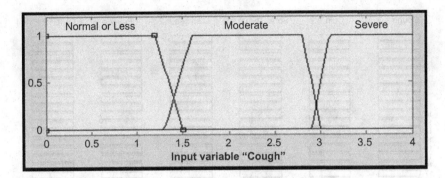

FIGURE 1.5 Membership function of cough.

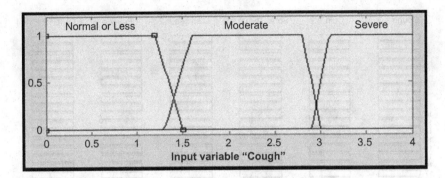

FIGURE 1.6 Rule editor.

exhibiting and writing rules. Figure 1.6 displays the rules of fis 1, i.e. 243 rules for cough, cold, sore throat, diarrhea, fever. The formula to make the rules is [10].

$$\text{Rules} = M^i$$

$$M = \text{Membership Function}$$

$$i = \text{Input Parameters}$$

1.4.4 FUZZIFICATION AND DEFUZZIFICATION

Fuzzification and defuzzification [50] play a very important role in the fuzzy inference system. Fuzzification is the process of converting the crisp input into a fuzzified output, and defuzzification converts the fuzzified value into a crisp value. In the proposed system, the centroid method is used for defuzzification.

1.4.5 RULE VIEWER

Rule viewer is put forth for analysis of the fuzzy inference system. Work out this examination as a signal for confirmation, e.g. the unit membership function's

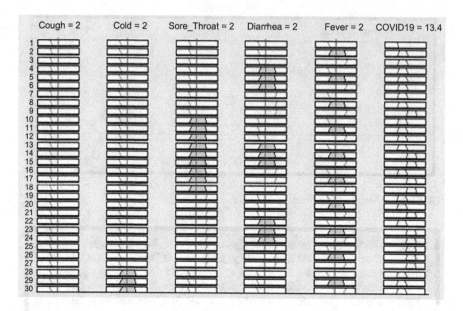

FIGURE 1.7 Rule viewer.

appearance involves the result. The rule viewer unveils the information of the utter fuzzy inference method. In accretion, the menu bar and status line are personal objects. The specific input value can be entered in the text field made available at the lower right location while framing. There is a text field in the bottom right, where we can submit a specific input value. Figure 1.7 shows a rule viewer of the projected organization. It shows the result of the whole fuzzy system. On the left plane at the crest, we get = 13.4 (defuzzified values) which means the person is normally found by the system.

1.4.6 Surface Viewer

For evaluating the dependence of any one output on one or two of the inputs surface viewer is used; for the FIS, it spawns and devises an output surface plot. Figure 1.8 shows the surface plot of the disease between two symptoms, cold and cough. Input is represented by blue color and output is represented by yellow color.

1.4.7 Graphical User Interface

In the current scenario, we developed a graphical user interface (GUI) to make it simple and easy to use for the initial self-screening of COVID-19 as shown in Figure 1.9. In Figure 1.9(a), there are 15 input parameters for the diagnosis of COVID-19 including cough, cold, sore throat, diarrhea, fever, body aches, fatigue, headache, difficulty in breathing, travel in the past 14 days, travel history in a COVID-infected area, direct contact or taking care of COVID patients, age, other medical conditions, and

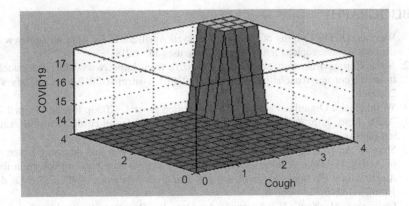

FIGURE 1.8 Surface view of cold and cough.

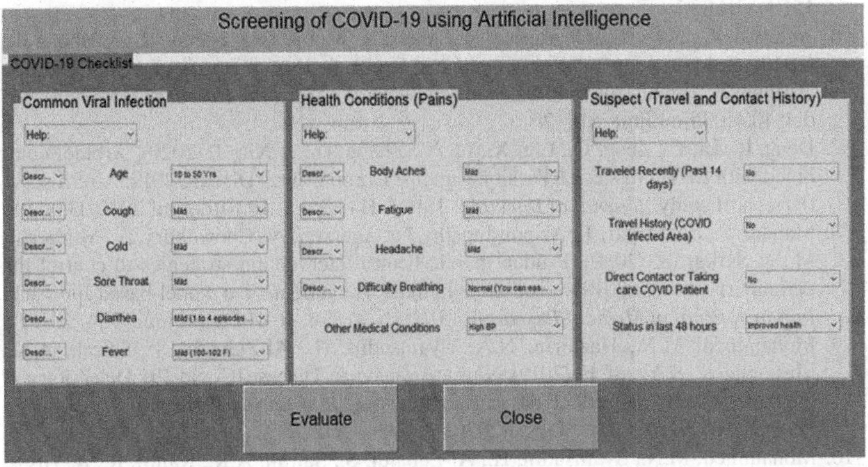

FIGURE 1.9 GUI screening assessment of COVID-19 as per input parameters.

status in the last 48 hours. The graphical user interface is linked with FIS files so that the users can easily interface with the system to give inputs and get results according to the rules.

1.5 CONCLUSION

COVID-19 is a widely spread harmful virus nowadays. So, there is a dire need for an initial screening system. This research introduces a hierarchical fuzzy expert system which is based on fuzzy rules. Pertaining to these it decides whether a person is healthy or if there is a need to seek consultation with a doctor. The graphical user interface has also been designed so that users or their care takers can easily access it.

BIBLIOGRAPHY

1. Wessner, D. The Origins of Viruses. Retrieved April 20, 2020, from https://www.nat ure.com/scitable/topicpage/the-origins-of-viruses-14398218/.
2. Novel Coronavirus. Retrieved April 20, 2020, from https://www.who.int/docs/defau lt-source/coronaviruse/situation-reports/20200121-sitrep-1-2019-ncov.pdf?sfvrsn= 20a99c10_4.
3. World Health Organization. COVID-19 situation reports. (2020). Retrieved April 20, 2020, from https://www.who.int/emergencies/diseases/novel-coronavirus-2019/situat ion-reports.
4. Sharma, A., Khosla, A., Khosla, M., & Rao, Y. (2018). Fast and Accurate Diagnosis of Autism (FADA): A novel hierarchical fuzzy system based autism detection tool. *Australasian Physical & Engineering Sciences in Medicine*, 41(3), 757–772. doi: 10.1007/s13246-018-0666-3.
5. Chatterjee, P., Nagi, N., Agarwal, A., Das, B., Banerjee, S., Sarkar, S., ... & Gangakhedkar, R.R. (2020). The 2019 novel coronavirus disease (COVID-19) pandemic: A review of the current evidence. *The Indian Journal of Medical Research*, 151(1), 71–83.
6. Agarwal, A., Nagi, N., Chatterjee, P., Sarkar, S., Mourya, D., Sahay, R., & Bhatia, R. (2020). Guidance for building a dedicated health facility to contain the spread of the 2019 novel coronavirus outbreak. *Indian Journal of Medical Research*, 151(1), 41–49. doi: 10.4103/ijmr.ijmr_518_20.
7. Deng, L., Li, C., Zeng, Q., Liu, X., Li, X., Zhang, H., & Xia, J. (2020). Arbidol combined with LPV/r versus LPV/r alone against Corona Virus Disease 2019: A retrospective cohort study. *Journal of Infection*, 151(1), 11–32. doi: 10.1016/j.jinf.2020.03.002.
8. Mandal, S., Bhatnagar, T., Arinaminpathy, N., Agarwal, A., Chowdhury, A., Murhekar, M., & Sarkar, S. (2020). Prudent public health intervention strategies to control the coronavirus disease 2019 transmission in India: A mathematical model-based approach. *Indian Journal of Medical Research*, 151(1), 1–10. doi: 10.4103/ijmr.ijmr_504_20.
9. Mohammed, M.N., Hazairin, N.A., Syamsudin, H., Al-Zubaidi, S., Sairah, A.K., Mustapha, S., & Yusuf, E. (2019). Novel Coronavirus Disease (Covid-19): Detection and diagnosis system using IOT based smart glasses. *International Journal of Advanced Science and Technology*, 29(7s), (2020), pp. 954–960.
10. Mohammed, M.N., Syamsudin, H., Al-Zubaidi, S., Sairah, A.K., Ramli, R., & Yusuf, E. (2020). Novel Covid-19 detection and diagnosis system using IOT based smart helmet. *International Journal of Psychosocial Rehabilitation*, 24(7), 2296–2303.
11. Pham, Q.-V., Nguyen, D.C., Huynh-The, T., Hwang, W.-J., & Pathirana, P.N. (2020). Artificial Intelligence (AI) and big data for coronavirus (COVID-19) pandemic: A survey on the state-of-the-arts. *IEEE Access*, 1–1. doi: 10.1109/access.2020.3009328.
12. Waheed, A., Goyal, M., Gupta, D., Khanna, A., Al-Turjman, F., & Pinheiro, P.R. (2020). CovidGAN: Data augmentation using auxiliary classifier GAN for improved Covid-19 detection. *IEEE Access,* 1–1. doi: 10.1109/access.2020.2994762.
13. Abdel-Basset, M., Mohamed, R., Elhoseny, M., Chakrabortty, R.K., & Ryan, M. (2020). A hybrid COVID-19 detection model using an improved marine predators algorithm and a ranking-based diversity reduction strategy. *IEEE Access*, 1–1. doi: 10.1109/ access.2020.2990893.
14. Symptoms of Coronavirus. (2019). Retrieved from https://www.cdc.gov/coronavirus /2019 ncov/symptoms-testing/symptoms.html.
15. Kumar, B.N., Chauhan, R.P., & Dahiya, N. (2016). Detection of Glaucoma using image processing techniques: A review. 2016 International Conference on Microelectronics, Computing and Communications (MicroCom). doi: 10.1109/microcom.2016.7522515.

16. Wang, N., Liu, H., & Xu, C. (2020). Deep learning for the detection of COVID-19 using transfer learning and model integration. 2020 IEEE 10th International Conference on Electronics Information and Emergency Communication (ICEIEC). doi: 10.1109/iceiec49280.2020.9152329.

17. Dong, D., Tang, Z., Wang, S., Hui, H., Gong, L., Lu, Y., & Li, H. (2020). The role of imaging in the detection and management of COVID-19: A review. *IEEE Reviews in Biomedical Engineering*, 1–1.

18. Chamola, V., Hassija, V., Gupta, V., & Guizani, M. (2020). A comprehensive review of the COVID-19 pandemic and the role of IOT, drones, AI, Blockchain and 5G in managing its impact. *IEEE Access*, 1–1. doi: 10.1109/access.2020.2992341.

19. Alsaeedy, A.A.R., & Chong, E. (2020). Detecting regions at risk for spreading COVID-19 using existing cellular wireless network functionalities. *IEEE Open Journal of Engineering in Medicine and Biology*, 1–1. doi: 10.1109/ojemb.2020.3002447.

20. Hossain, M.S., Muhammad, G., & Guizani, N. (2020). Explainable AI and mass surveillance system-based healthcare framework to combat COVID-I9 like pandemics. *IEEE Network*, 34(4), 126–132. doi: 10.1109/MNET.011.2000458.

21. Hakak, S., Khan, W.Z., Imran, M., Choo, K.R., & Shoaib, M., (2020). Have you been a victim of COVID-19-related cyber incidents? survey, taxonomy, and mitigation strategies. *IEEE Access*, 8, 124134–124144. doi: 10.1109/ACCESS.2020.3006172.

22. Michailidis, E.T., Potirakis, S.M., & Kanatas, A.G. (2020). AI-Inspired Non-Terrestrial networks for IoT: Review on enabling technologies and applications. *IoT*, 1(1), 21–48.

23. Kunhimangalam, R., Ovallath, S., & Joseph, P.K.. (2013). A novel fuzzy expert system for the identification of severity of carpal tunnel syndrome. *BioMed Research International*, 2013, Article ID 846780, 12 pages. http://dx.doi.org/10.1155/2013/846780

24. Ephzibah, E.P., & Sundarapandian, V. (2012). A fuzzy rule based expert system for effective heart disease diagnosis. In: Meghanathan, N., Chaki, N., Nagamalai, D. (eds) Advances in Computer Science and Information Technology. Computer Science and Engineering. CCSIT 2012. Lecture Notes of the Institute for Computer Sciences, Social Informatics and Telecommunications Engineering, vol 85. Springer, Berlin, Heidelberg. https://doi.org/10.1007/978-3-642-27308-7_20

25. Painuli, D., Mishra, D., Bhardwaj, S., & Aggarwal, M. (2020). Fuzzy rule based system to predict COVID19 - a deadly virus. *International Journal of Management and Humanities (IJMH)*, 4(8), ISSN: 2394-0913.

26. Nema, B.M., Makki Mohialden, Y., Mahmood Hussien, N., & Ali Hussein, N. (2020). COVID-19 knowledge-based system for diagnosis in Iraq using IoT environment. *Indonesian Journal of Electrical Engineering and Computer Science*, 15(3), xx–xx. ISSN: 2502-4752, doi: 10.11591/ijeecs.v15.i3.

27. Sagir, A.M., & Sathasivam, S. (2017). Intelligence system based classification approach for medical disease diagnosis. AIP Conference Proceedings. doi: 10.1063/1.4995879.

28. Fatima, S.A., Hussain, N., Balouch,A., Rustam, I., Saleem, Mohd, & Asif, M. (2020). IoT enabled smart monitoring of coronavirus empowered with fuzzy inference system. *International Journal of Advance Research, Ideas and Innovations in Technology*, 6(1).

29. Memish, Z.A., Cotten, M., Meyer, B., Watson, S.J., Alsahafi, A.J., Al Rabeeah, A.A., Corman, V.M., Sieberg, A., Makhdoom, H.Q., Assiri, A., & Al Masri, M. (2014). Human infection with MERS coronavirus after exposure to infected camels, Saudi Arabia, 2013. *Emerging Infectious Diseases*, 20(6), 1012.

30. Müller, M.A., Corman, V.M., Jores, J., Meyer, B., Younan, M., Liljander, A., Bosch, B.J., Lattwein, E., Hilali, M., Musa, B.E., & Bornstein, S. (2014). MERS coronavirus neutralizing antibodies in camels, Eastern Africa, 1983–1997. *Emerging Infectious Diseases*, 20(12), 2093.

31. Lai, C.C., Shih, T.P., Ko, W.C., Tang, H.J., & Hsueh P.R. (2020). Severe acute respiratory syndrome coronavirus 2 (sars-cov2) and coronavirus disease-2019 (COVID-19), the epidemic and the challenges. *International Journal of Antimicrobial Agents*, 55(3), 105924.

32. Sikchi, S.S., Sikchi, S., & Ali, M.S. (2013). Fuzzy expert systems (FES) for medical diagnosis. *International Journal of Computer Applications*, 63(11), 7–16.

33. Chatterjee, P. et al. (2020). The 2019 novel coronavirus disease (Covid-19) pandemic: A review of the current evidence. *Indian Journal of Medical Research, Supplement*, 151(2–3), 147–159. Indian Council of Medical Research. doi: 10.4103/ijmr.IJMR_519_20.

34. Pascarella, G. et al. (2020). COVID-19 diagnosis and management: A comprehensive review. *Journal of Internal Medicine*. doi: 10.1111/joim.13091.

35. West, C.P., Montori, V.M., & Sampathkumar, P. (2020). COVID-19 testing: The threat of false-negative results. Mayo Clinic Proceedings. Elsevier Ltd. doi: 10.1016/j.mayocp.2020.04.004.

36. Wang, D. et al. (2020). Clinical characteristics of 138 hospitalized patients with 2019 novel coronavirus-infected pneumonia in Wuhan, China. *JAMA – Journal of the American Medical Association*, 323(11), 1061–1069. doi: 10.1001/jama.2020.1585.

37. Huang, C. et al. (2020). Clinical features of patients infected with 2019 novel coronavirus in Wuhan, China. *The Lancet*, 395(10223), 497–506. doi: 10.1016/S0140-6736(20)30183-5.

38. Luz, E., Silva, P.L., Silva, R., Silva, L., Moreira, G., & Menotti, D. (2020). Towards an effective and efficient deep learning model for COVID-19 patterns detection in x-ray images. Retrieved from http://arxiv.org/abs/2004.05717.

39. Salehi, S., Abedi, A., Balakrishnan, S., & Gholamrezanezhad, A. (2020). Coronavirus disease 2019 (COVID-19): A systematic review of imaging findings in 919 patients. *American Journal of Roentgenology*, 215, 1–7.

40. Hosseiny, M., Kooraki, S., Gholamrezanezhad, A., Reddy, S., & Myers, L. (2020). Radiology perspective of coronavirus disease 2019 (COVID-19): Lessons from severe acute respiratory syndrome and middle east respiratory syndrome. *American Journal of Roentgenology*, 214(5), 1078–1082.

41. Chung, M., Bernheim, A., Mei, X., Zhang, N., Huang, M., Zeng, X., Cui, J., Xu, W., Yang, Y., Fayad, Z.A., Jacobi, A., Li, K., Li, S., & Shan, H., (2020). CT imaging features of 2019 novel coronavirus (2019-nCoV). *Radiology*, 295(1), 202–207.

42. WHO. (2020). Director-General's Opening Remarks at the Media Briefing on COVID-19–11 March 2020. Retrieved from https://www.who. int/dg/speeches/detail/who-director-general-s-op

43. Coronavirus Disease 2019 (COVID-19). (2020). Retrieved from https://www.cdc.gov/coronavirus/2019-ncov/need-extra-precautions/ people-at-higher-risk.html

44. Medicine, J.H.U. (2020). Coronavirus COVID-19 Global Cases by the Center for Systems Science and Engineering (CSSE) at Johns Hopkins University (JHU). Retrieved from https://coronavirus.jhu.edu/map.html

45. Wang, W., Xu, Y., Gao, R., Lu, R., Han, K., Wu, G., & Tan, W., (2020). Detection of SARS-CoV-2 in different types of clinical specimens. *Jama*, 323(18), 1843–1844.

46. Yang, T., Wang, Y.-C., Shen, C.-F., & Cheng, C.-M., (2020). *Point-of-Care RNA Based Diagnostic Device for COVID-19*. Basel, Switzerland: Multidisciplinary Digital Publishing Institute.

47. News, A.J.. (2020). India's Poor Testing Rate May Have Masked Coronavirus Cases. Retrieved from https://www.aljazeera.com/news/2020/03/india-poor-testing-rate-masked-coronaviruscases-200318040314568.html.

48. Wetsman, N.. (2020). Coronavirus Testing Shouldn't Be This Complicated. Retrieved from https://www.theverge.com/2020/3/17/21184015/coronavirus-testing-pcr-diagnostic-point-of-care-cdc-techonology.

49. Solanki, A., & Kumar, E. (2017). Design and development of web enabled fuzzy expert system using rule advancement strategy. *International Journal of Intelligent Systems Design and Computing*, 1(1/2), 3–27.

50. Solanki, A., & Kumar, E. (2013). A novel technique for rule advancement in fuzzy expert system using einstein sum. In: Confluence 2013: the next generation information technology summit (4th international conference), pp. 62–68.

2 An Insight into the Potential Role of Artificial Intelligence in Bioinformatics

Divya Bhatia and Deepak Malik

CONTENTS

2.1 INTRODUCTION

Bioinformatics, a continuously emerging field, employs computational and biological information to answer biological queries. Bioinformatics mainly deals with the organization and study of biological data. Bioinformatics is an interdisciplinary field that includes mathematical models, statistical approaches, and algorithms to solve numerous biological problems [1]. Bioinformatics tools have significantly helped in deciphering and interpreting the continuously increasing biological data produced through genome sequencing projects. The two most significant bioinformatics functions are (1) the development of software tools and algorithms (2) the analysis and interpretation of biological data using developed software tools and algorithms. The applications of bioinformatics methods include the organization and performance of DNA sequences and protein structures, gene prediction evolutionary studies,

DOI: 10.1201/9781003126164-2

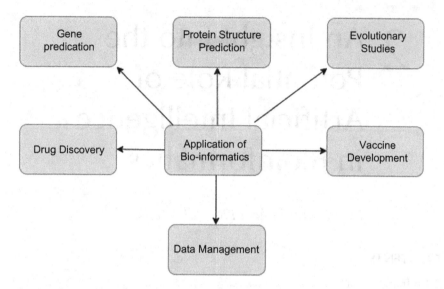

FIGURE 2.1 Major applications of bioinformatics.

protein structure prediction, and vaccine development, as shown in Figure 2.1. The bioinformatics methods have recently been employed for modeling, in silico experiments, and the simulation of biological systems. Bioinformatics helps in accessing and interpreting the data and studying the mechanism of biological processes [2]. Several biological problems have been solved using bioinformatics-based methods [3]. The advances in genomic studies have drastically affected the existing scenario. The generation of large amounts of biological data has created the need to develop tools and methods for sequence analysis and the modeling of biological systems and processes. Computation-based bioinformatics methods have augmented the speed of acquiring and exploiting data and significantly influenced biological research. Thus, these methods have become imperative in biology to handle the continuously increasing massive amount of data.

Various options are presented to understand the biological data and associated challenges. Artificial intelligence is a technique for developing a system with intelligent thinking and simulation of human beings [4]. Several advances in bioinformatics and artificial intelligence have been observed in the last decade. Artificial intelligence has applications in structural bioinformatics and can be employed to design novel compounds against various neurological and genetic diseases by using in silico techniques [5][6][7]. Bioinformatics-based tools assist in the analysis of biological data to find a logical conclusion.

2.2 ARTIFICIAL INTELLIGENCE

Artificial intelligence is related to multiple disciplines like computer science, life science, psychology, linguistics, mathematics, and engineering. Artificial intelligence

(AI) is an approach to designing a computer, robot, or software with intelligent thinking power similar to humans. AI is based on the study of human brain thinking, learning ability, decision-making, and the ability to solve problems. These research-based outcomes are used to accomplish AI. Artificial intelligence includes the development of intelligent software. AI's primary function is developing methods or algorithms with human intelligence, such as analysis, knowledge, and problem-solving.

Artificial intelligence is the machine's ability to think, create, and execute the work intelligently based on its training [8]. The AI theory is based on the concept that human thinking and reasoning can be inserted into machines. AI exploits different statistics tools, knowledge engineering, and neural network approaches to execute various tasks like knowledge retrieval, speech recognition, planning, thinking, reasoning, and manipulation. Human intelligence cannot be precisely embedded in machines yet, but several products based on AI concepts have been designed for use.

2.2.1 OBJECTIVES OF AI

i) *Intelligent systems development*: the systems with smart learning, thinking, performance, and the ability to explain and demonstrate.
ii) *To implement human intelligence in machines*: develop machines to think, learn, and perform like humans.

AI-based systems are designed with a target of attaining the efficient utilization of knowledge. The developed system should be easily perceivable by the people and modifiable to correct errors. AI techniques increase the processing speed of associated complex programs. AI is categorized into two categories, generalized and applied AI. The generalized AI includes designing a machine with human-like performance ability, and applied AI includes the simulation of human-like thinking and expressions. Five primary attributes of AI are shown in Figure 2.2.

(a) *Expert system*: this involves an understanding of human intelligence and machines to impart the ability to give suggestions, demonstrations, and explanations of its conclusions. This system is based on a database with excellent knowledge about the specific area. The development of this kind of program is known as knowledge engineering. An expert system-based program contains the understanding of human experts, and the transfer

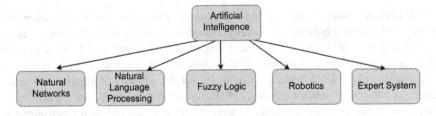

FIGURE 2.2 Major attributes of AI.

TABLE 2.1

The Role of Different Components of AI in Bioinformatics

Categories	Biomedical Imaging	Biomedical Signal Processing
Neural networks	Segmentation	Classification
NLP	Brain decoding	Brain decoding
Fuzzy logic	Recognition	Brain decoding
Robotics	Brain decoding	Brain decoding
Expert system	Recognition	Brain decoding

Here it shows where we apply expert systems, robotics, neural networks, NLP, etc., to predict brain decoding, recognition, segmentation applications of biosignal processing, and biomedical imaging.

of knowledge takes place through the interaction between experts and the system. However, the expert system development is still in the initial stage. Some presently available expert systems are MYCIN, TURNX, and *PROSPECTOR* designed for training and troubleshooting.

(b) *Fuzzy logic*: fuzzy logic is a tool to control industrial operations in diagnostics, electronics, and household instruments. Fuzzy logic can handle those critical situations that true/false logic cannot. Fuzzy logic is a mathematical-based problem-solving approach for processing inaccurate data for complex and non-linear processes. Approximate values in between yes/no, true/false can be generated and executed. So, this is an effort to develop a computer with human ways of thinking.

(c) *Artificial neural network (ANN)*: **the** ANN concept is based on the idea of mimicking humans to execute complex tasks that regular computer systems cannot perform. Alan Turing in 1948 first proposed the idea. Neural networks are designed not only to simulate the human brain but also to get the information similar to brain. An ANN is a network of the neuron that can accomplish complicated tasks. ANN has applications in system identification, operational control, decision-making, pattern recognition, and data processing (Table 2.1).

2.3   NEED FOR INTEGRATION OF AI AND BIOINFORMATICS

DNA sequence analysis has opened various opportunities for computer scientists owing to digital information availability. However, there are many challenges associated with this field, like (1) the identification and annotation of DNA segments with specific biological functions like protein-coding sequences, and regulatory sequences, and (2) DNA sequence alignment for the analysis of identity and similarity. Advances in omics technology, high-throughput methods, and system biology have generated huge amounts of available and accessible data electronically. However, the major challenge is processing and analyzing these data at the speed of data generation. This condition, considered "big data," has generated the need to design new management tools to analyze and annotate big data [9]. For the best possible utilization of big data, AI can assist in different ways: enabling the efficient application of algorithms and executing multiple simple operations over large complex spaces of biological data.

Another major challenge in bioinformatics is identifying differentially expressed genes that exhibit different expression levels between two conditions: healthy/pathological, experimental conditions, developmental stages, etc. [10]. Some researchers are working on identifying entire gene networks and differentially expressed metabolic pathways [11]. Another challenging task is the identification of alternative transcripts.

There is a dire need to develop novel methods of gene extraction and protein network mapping from continuously expanding gene and protein databases. There is very little knowledge about the existing techniques of gene expression analysis and proteomic analysis like clustering, correlation estimation, and self-organizing maps, which can directly lead to reverse engineering. The amalgamation of AI and bioinformatics has helped develop algorithms and software for gene prediction, in silico drug designing, protein interaction studies, and next-generation sequencing. AI-based bioinformatics mainly utilizes neural networks, decision trees, hybrid methods, and genetic algorithms to solve different biological issues.

AI-based methods have significant potential for bioinformatics' current and future developments and can exhibit an intense effect on the pharmaceutical and biotechnology industry. Recently, applied AI has been exploited to develop personalized medicine by the genome data analysis of patients. AI-based algorithms have the advantage of efficient data ability compared to algorithms without AI. AI-based algorithms such as linear regression, logistic regression, and k-nearest neighbors algorithms in bioinformatics may increase computational biology's effectiveness in overcoming the existing challenges [12].

AI has impacted various bioinformatics applications by facilitating the parsing of genome sequences, sequence alignment, structure-function analysis, and protein interaction-related study [13]. AI also assists in advanced bioinformatics applications such as in silico research, simulations of systems, complex systems study, and computer-aided drug design. Machine learning and automated reasoning based on advanced algorithms have analyzed complex systems. Recent advances in AI and bioinformatics like high-throughput screening, epidemiology, and immune-informatics have contributed to improved simulation and significantly impacted human health [14]. Continuously mutating microbes and viruses has made vaccine development a challenging task. Recently, with the advent of technology, the ability to screen the potential target has increased significantly.

2.4 APPLICATION OF AI IN BIOINFORMATICS

The bond between artificial intelligence and bioinformatics is well established. AI has a broad range of bioinformatics applications, and its primary functions are visual cognizance, voice recognition, decision-making, and translation between languages.

Biomedical applications involve development, medical image analysis, diagnostics, examination, and robotic surgery [15]. AI deals with the problems related to the effects of variable parameters in population genetics. Linear analysis algorithms and indexation techniques are used to face the challenges.

Biological databases contain a considerable amount of data that is continuously increasing. The interpretation and mining of these vast data are big challenges for the researcher. AI-based methods can assist in bioinformatics for data analysis to get a logical conclusion. The combination of AI and bioinformatics may result in the efficient simulation of various models, the analysis of genome sequences, drug designing, virtual screening, and gene prediction [16]. The primary function of AI in bioinformatics analysis is based on pattern matching and knowledge-based learning systems.

The first success was "Protean," a system developed to analyze nuclear magnetic resonance data to predict 3D structure [17]. Then the practical machine learning application Meta-Dendral was developed. Presently, the analysis of mass spectrum data related to metabolites and peptide is an essential requirement in biology. So, the application and assistance of artificial intelligence can be observed everywhere in bioinformatics. Some of the major applications of AI in bioinformatics are discussed below.

2.4.1 DATA AND KNOWLEDGE MANAGEMENT

Data generated from various molecular biology and high throughput-based methods after automated interpretation are deposited in several databases. These data's maintenance and organization are challenging due to the continuous increase in data at different quality levels. The integration and maintenance of the data from various sources are a significant concern and require recurrent schema and updating processes. These issues are solved using a graphical representation of heterogeneous data. The other challenge is to support and control the integration and interaction of data. Semantic web technologies like RDF, SPARQL, and OWL from the W3C have supported the linked Open Data initiative [18]. These technologies assist in the integration of bioinformatics data.

Genome assembly is considered a very complicated task because it requires many sequences to join together. Assembly programs should generate one contig for every chromosome of the genome. However, many gaps are left after the joining of contig due to the complexity of the genome. Another major problem is due to the presence of repeats. Repeats present may be tandem or interspersed throughout the genome. The assembler cannot differentiate between numbers of repetitions, which may lead to error. Other genome analysis problems may be due to contaminants from other organisms, sequencing error due to the presence of homopolymer, PCR product error like chimera, polyploid genomes, etc. In bioinformatics, machine learning-based applications have to deal with some challenges like sample size, multidimensionality, and complex relations between data. Ensemble learning is used to reach a robust conclusion. It has been used in numerous tasks like gene expression, regulatory elements prediction, structure prediction of proteins, and interaction study [19].

2.4.2 Information Extraction in Biological Literature

Previously, biological information was retrieved manually with the help of two specialists, an expert curator for the extraction of data from publications in a computer-friendly format and other experts in text mining related to biological challenges. Manual execution is a tedious and time-consuming task that cannot match the speed of continuously increasing biological data. Regulon DB, a curated database, showed that information could be mined using natural language processing. BioGRID database and BioCreative have jointly produced the compilation of 120 full-text articles called the BioC-BioGRID corpus. This collection contains a description of biological entities, protein, and gene interaction. OntoGene has been used in RegulonDB [20]. The OntoGene system is a text mining system developed to find different species and interaction studies between genes, proteins, drugs, and diseases. It also performs the identification of segments, sentence splitting, stemming, and syntactic analysis. This system employs a rule-based approach for recognition and HMM-based learning for disambiguation.

2.4.3 Gene and Noncoding RNA Prediction

The first step of genome annotation is the identification of protein/RNA coding and noncoding segments. In prokaryotes, protein genes are present as a single segment without intron, making gene prediction relatively easy. However, the main issue is related to predicting the ribosomal frameshift event, which can be handled using HMM prediction. But it isn't easy to get experimentally verified data, limiting the applications of model learning. The ribosome profiling technique (Ribo-Seq) has been developed to map the specific ribosomal location on translating mRNA [21].

Initially, algorithms developed for gene prediction were focused only on DNA sequence due to limited RNA sequencing methods. For the transcribed sequences and populations' genetic studies, expressed sequence tags (EST) are used. RNA-Seq technology and mass spectrometry data have revolutionized the process of gene discovery. GeneMark, GeneID, Glimmer HMM, AUGUSTUS, and SNAP are important gene finders that function by identifying specific parameters in DNA sequences. They also predict the functions of closely associated sequences such as regulatory regions. These modeling systems are mainly HMM-based.

2.4.4 Protein Structure Prediction

From the last few decades, intensive in vitro research has been conducted to find the structure of proteins based on different lab-based experiments like nuclear magnetic resonance and X-ray crystallography. But these methods are labor-intensive, expensive, and require years of research. The prediction of 3D protein structure is one of the major bioinformatics issues and still needs intensive study. The basic approach for complete 3D structure prediction is shown in Figure 2.3. AI-based methods can overcome the limitations of these methods and may prove to be a better alternative to the traditional experiment-based approach. Normally, linear programming,

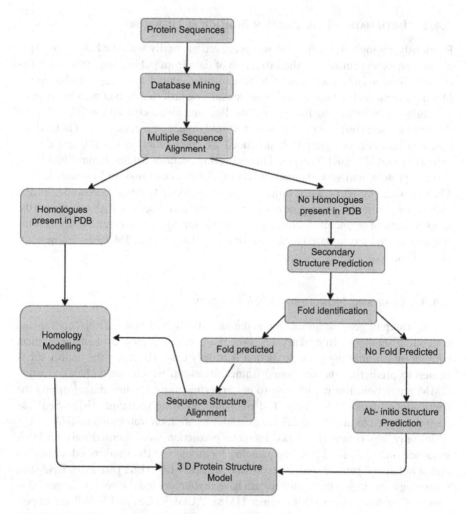

FIGURE 2.3 Basic strategy for 3D structure prediction.

quadratic programming, weighted partial MaxSAT, and graphical model optimization-based approaches are used to solve associated issues.

Secondary structure prediction is one of the major bioinformatics challenges. The complete elucidation of 3D protein structure and protein folding kinetics study begins with a secondary structure prediction (SSP). Various machine learning methods have been used to normalize secondary systems, take corrective procedures, and propose standard datasets.

Normally, a prediction is carried out in two steps: (1) primary prediction of class, (2) standardization of the forecast over the sequences to find similarities between different positions. Initially, a similar protein recognition study was conducted based on the concept that all proteins are evolutionarily related rather than using protein sequences. This concept has helped improve the prediction score and find segments with high confidence predictions. After collecting protein similarity data, proteins can be classified in the form of a profile HMM, based on multiple sequence alignment. Based on ensemble strategy and neural networks, three important prediction servers, PSIPRED, Distill, and Jared [22], have been developed. The Protein Data Bank contains non-redundant data related to protein chains with known secondary and 3D structures.

In 2014, a bidirectional recursive neural network was proposed to assign the secondary structure to regions with sequence similarity. However, the prediction of structure using already present known systems as the template does not provide folding information. We have to rely on this strategy as there is a huge gap between total protein sequences and known protein structures available. It is very difficult to predict the complete 3D structure of the protein, so it is essential to analyze the efficacy of prediction methods critically. Even deep learning methods based on restricted Boltzmann machines did not show any significant improvement in prediction.

In contrast, a recently developed deep learning network has produced improved results [23]. Deep convolutional neural field (Deep CNF), available on server RaptorX, is based on the combination of conditional random field and a deep convolutional neural network. It considers the correlation between secondary structure at various positions and long-range correlation in sequences. Long-range correlation consideration is essential for the improvement in secondary structure predictions. Similarly, other architecture, long short-term memory bidirectional recurrent neural network (LSTM-BRNN) available on Spider3, has also shown good results. Some important features like intrinsically disordered, exposed regions, capping regions, and proline conformation need to be considered to predict SSP and hairpin structures for folding kinetics [24].

The prediction of dihedral angles or torsion angles and the prognosis of contacts (<8 Å) for the elucidation of protein backbone structure is also a major challenge in structural bioinformatics. SPINE-X has shown promising results in predicting the dihedral angle [24]. With the help of deep learning, it is easy to integrate multiple parameters and a combination of recurrent networks.

The most complex problem is the generation of 3D protein structures. Contact map information is used in machine learning to predict the 3D design [25]. The Confold method utilizes the approach already used to predict the torsion angle to frame a rough 3D model from secondary structure contacts. In this method, a distance geometry algorithm is used to generate a rough model that can be further refined [70]. Most of the methods depend on multiple sequence alignment and correlation between different positions to predict contacts [26]. References can be expected from various alignments by using machine learning methods. Normally, a combination of two forms, such as eliminating unlikely contact and the addition or predictions of missed native patterns of communications, is used [27]. Recently, the other strategy involves the partition of torsion angles into specific positions before prediction. This discretion is based on the Ramachandran plot that some strong preference regions are present for possible torsion angles. According to Anfinsen's hypothesis, proteins' folding information is present in their respective primary amino acid sequence. Based on this concept, various simulation programs analyze protein conformation and search for the lowest free energy, i.e., the native state. Recently, the protein structure annotations (PSA) approach, which involves intermediate steps prediction, has simplified the task of complete 3D structure prediction [28]. The most frequently used PSA are the secondary structure, solvent accessibility, and contact maps.

Despite the utmost importance of PSA in structure prediction, its role is less clear than other approaches [29]. Protein intrinsic disorder with structural and functional importance has also been predicted by machine learning (ML) and deep learning methods (DL) similar to the methods used for 1D PSA [25][30].

Since the beginning of the third generation of predictors, numerous deep learning algorithms [28], ML methods such as k-nearest neighbors, linear regression, hidden Markov models, support vector machines (SVM), and support vector regression [92] have been developed and used for PSA prediction. Depending on the number of input sequences, a newly developed predictor, NetSurfP-2.0, either uses HHblits or MMsEqs [31][32][33].

All proposed methods are based on the coarse-grained models by decreasing the number of atoms taken into account. The prediction of a protein with a large number of particles is the most difficult task. Different models like the CABS model, SICHO model, and Rosetta have been developed, which can be used in various applications in structural biology. The CABS model is analogous to the continuous dynamic approaches and can predict different secondary structures [34]. Numerous lattice models have been developed to predict the folding of native proteins concerning secondary structures. The most important challenge is identifying symmetries responsible for identical conformations. Numerous HP model base variants are available for working in the hexagonal lattice [35]. Another strategy is robot path planning based, where amino acids are added sequentially to form peptide chains on the lattice [36].

In another study, an enhanced energy function is proposed for square lattice models considering the interactions except for hydrogen–hydrogen [37]. One study has developed mixed-integer programming and an exact algorithm and two heuristic algorithms [38]. Homology modeling and protein threading methods depend on the alignment of a protein sequence on a known structure. In homology modeling, a homologous protein of available form is considered a template. However, protein threading is based on the concept that the collection of core templates is present, consisting of sequences of different lengths. The most suitable alignment is generated by optimizing the score between the arrangements and the template. It is difficult to find the optimum alignment because of gaps in different lengths and interactions between neighboring amino acids. Most of the alignment methods utilize the dynamic programming approach with some additional parameters like amino acid preferences, mutations, and solvent accessibility. RaptorX is a protein threading program based on regression trees to calculate each alignment's scoring function [39]. This model utilizes a neural network to rank and measure the quality of alignment.

The protein docking involves the search for minimum free energy conformations of a receptor-ligand complex. Template-based modeling is mainly employed for the prediction of the conformation of complexes [40]. Based on the folding kinetics of known protein structures, protein folding pathways are predicted and represented by roadmaps.

The main aim of computational protein design is to design new sequences with a fold similar to the target protein structure. Rational protein design utilizes rotamers libraries and is based on complex energy functions. Logical protein designs utilize rotamers libraries and can be considered as an optimization based on energy functions. Artificial intelligence and bioinformatics together can help to overcome the limitations in the existing strategy [41]. Recently, a deep learning-based model Alpha Fold has been developed to model protein structure based on genetic sequence only. This system is superior to the previous modeling system, and work accuracy is equivalent to existing systems. This research approach is based on a deep neural network and three sets of algorithms. . One algorithm is trained to predict the distance and angles between amino acids and the other two to predict space distribution between amino acids.. These algorithms were designed to measure the accuracy of the proposed protein structure. Finally, the gradient descent technique was employed to increase the accuracy of predicted structures further.

2.4.5 EVOLUTIONARY STUDIES

Multiple sequence alignment (MSA) is one of the key issues associated with sequences in biology. It involves fixing the letters' position in various arrangements to show a maximum similarity among letters at a specific place. Sequence alignment (DNA/protein/RNA) is based on the concept that all species originated from common ancestors, and the sequences share some common characteristics. But the main problem in the alignment is that arrangements have become divergent due to mutation and insertion/deletion events. MSA and phylogeny are related; however, structural and functional properties may be considered.

Multiple sequence alignment helps to build the species' evolutionary tree and, conversely, the alignment of sequences. Multiple sequence alignment can also be used for sequence annotation through the identification of specific patterns (HMM profile) and grammatical inference studies (automata) [42]. Multiple alignments consider mutation events like point mutation and insertions/deletions, but complex events like duplication, inversions, or recombination events make the studies more complicated. So, further research is required to solve these complicated issues related to variability. Artificial intelligence can help find the shortest path in a graph joining the best possible aligned positions in a sequence, and the bound algorithm can be employed.

Recent approaches have been designed utilizing external disk space and multi-threaded computation [43]. Although various suboptimal techniques have already been developed, further development processes need to be continued for whole-genome sequences. Almost all these approaches work on the progressive strategy, which starts from pairwise alignment, and then results are combined to get the best possible order. In progressive alignment, each sequence is aligned with exactly two rows. Contrary to this, the main focus of multiple alignments is to find the circular order of sequences with the minimum sum. The progressive strategy, in combination with the iterative method for continuous refinement of alignment, is used. A genetic algorithm and simulated annealing algorithm are used for the advanced and iterative part, respectively [44]. Several aligners have been developed, but no model can perform in all the conditions (intrinsically unstructured regions, partial sequences, alternative 2D/3D structure regions). Normally, meta-methods are proposed that run multiple algorithms in parallel, and then the best alignment is selected. Recently, assisted multiple alignments have been employed for more relevant and precise alignments. AlexSys, an expert system designed for multiple sequence alignment, is based on rules that fit the aligned sequences or not [45].

Molecular phylogenetic studies attempt to find the actual ancestry of species during evolution by analyzing the phylogenetic tree. In addition to multiple sequence alignment, various other information sources are employed to construct, research, · and resolve phylogenetic trees. Various studies have attempted to achieve sequence alignment and phylogeny simultaneously. In a tree construction, different criteria have to be considered like a perfect phylogeny (character may be gained once and for all), persistent phylogeny (once attained and then lost at least once), caminsokal criterion (gained several times but never lost), etc. Constraint modeling frameworks can work with different criteria. Different models have been proposed on the basis of integer linear programming, the type of data to be processed like perfect phylogeny, maximum parsimony, perfect phylogeny from tumor multi sample sequences, persistent phylogeny, and incomplete perfect phylogenies on tumoral sequences. In recent years, the algorithm is not the only focus of evolutionary bioinformatics. A knowledge-based approach is gaining importance because it combines knowledge from various experts and integrates their results.

A model checking framework using computation tree logic (CTL) was developed to assess the properties of phylogenetic trees. A model checking framework using CTL on phylogenetic trees was developed. If the sources are not satisfied, not only validation but the possibility of counter-examples is important in a practical interactive context where intensive queries are raised to mine the trees. Further, through the use of stochastic logic, an attempt has been made at the treatment of quantitative information logics [46]. It is most important to check models of evolution and to calculate maximum likelihood estimated for trees. Other frameworks like answer set programming have been used. The supertree construction problem, encoded as an ASP model, works with an overlapping set of species for tree construction [47]. A web service interface, API, has been created in ASP for Treebase (relational database) for the management and exploration of data related to the phylogenetic relationship [48]. A toolkit using various reasoning systems (first order, answer set, and dedicated provers) has been developed for the alignment, consistency evaluation, and repair of taxonomies [49]. To improve the search efficacy, a divide and conquer approach may be used for the alignment of trees. Recently, a method based on splitting the set of characters into subsets has been used for the search for the best suited phylogeny and then "anchor" trees built on these subsets are used to search the whole tree [50].

2.4.6 DRUG DISCOVERY

The application of AI in drug discovery can overcome the limitations and uncertainties associated with classical drug development methods and will also minimize human intervention. Some major contributions of AI in drug development involve possible synthetic route prediction for drug-like molecules, pharmacokinetics properties, protein characteristics, drug-target interaction [51], and drug repurposing [52]. There are various available databases, such as ZINC, PubChem, Ligand Expo, KEGG, ChEMBL, DrugBank, STITCH, BindingDB, Super target, and PDB, that can be used to integrate information related to molecular pathways, crystal structures, drug targets, binding interactions, and biological activities. AI can be used to explore the information stored in databases to design polypharmacological agents (one disease multiple targets). The major advantage of the AI technique is that it has a minimum dependency on datasets. So, a network can be designed on the basis of drug molecule design.

The prediction of new pathways, personalized medicine, novel biomarkers, and target identification using omics technology are also major application areas of AI. An AI-based platform, IBM Watson, has successfully identified RNA-binding protein (RBPs) associated with the pathogenesis of the neurodegenerative disorder amyotrophic lateral sclerosis (ALS) [53]. The most important step in drug development is the synthesis of a selected drug molecule. Thus, AI can assist in the development of tools for synthetic pathways and can categorize the molecule on the basis of ease of synthesis [54]. A novel AI platform, 3N-MCTS, based on the combination of three different deep neural networks with Monte Carlo tree search, has been developed. This platform can prioritize, sort, and select optimum pathways for the synthesis of drug compounds. The intervention of AI can increase the efficacy of drug development; however, AI techniques need to be validated first before application in drug development.

Some algorithms have been designed to analyze the difference between the normal and diseased profiles. An AI-based tool SPiDER has been developed for natural products-based drug development [55]. The application of AI has assisted in making drug repurposing a realistic and attractive approach. Deep learning has been used in various in silico methods for the prediction of pharmacological properties of drugs and drug repurposing. The method was based on deep neural networks (DNNs), which are highly adaptive multilayer systems involving interacting artificial neurons for data transformation [56]. Next-generation AI-based drug development uses generative adversarial networks (GANs). DeepDDI is an AI-based platform for the better understanding of drug–drug interaction, mechanism of action, and prediction of better drugs without negative effects [57].

Biologically active target molecule identification and the exact prediction of their activity are challenging tasks. In the pharmaceutical industry, the development of new drugs is mainly based on these chemical databases and high-throughput screening (HTS) techniques, which have reduced the time and animal testing need. However, the major issue is to find structure–activity relationships (SAR) that help to predict activity on the basis of the physio-chemical properties of chemicals. It is assumed that similar molecules will exhibit similar activity; however, even some minor changes in structure can affect the activity. This concept is termed the activity cliff [58]. The two main approaches, structure-based drug design (SBDD) and ligand-based drug design (LBDD), are used for known and unknown structures respectively.

The biological activity of a drug is dose-dependent and can be illustrated by many parameters. The dose-dependent activity of a drug mainly depends on two factors, i.e., the target activity and associated toxicity. In pharmacology, it is measured in the form of absorption, distribution, metabolism, and excretion (ADME) properties. Initially, simple statistical-based methods like linear regression were employed for the structure–activity study of chemical compounds. In the last decade, machine learning techniques have significantly affected this field.

Various open source/commercial rule-based systems are available to study the structure–activity relation and for the prediction of activity in various areas like hepatotoxicity, mutagen compounds, and allergy-related compounds [59].

On the basis of available information, knowledge-based expert systems are mainly used to predict the associated toxicity. Toxtree is used for the evaluation of mutagenic chemicals' functions through a manually devised set of instructions [60]. DeepTox and PrOCTOR (a platform for the prediction of toxicity of new compounds) have been developed. In the prediction of xenobiotic metabolism, a lot of false-positives can be produced which makes it more challenging [61]. Some automated prediction methods are Bayesian methods, SVM, and deep learning. Random forests (RF), a decision tree, have been designed to predict the structure-activity relation and quality [62]. Recently, they have been utilized in toxicology prediction and protein-ligand docking studies [63].

2.4.7 VACCINE DEVELOPMENT

The application of AI in bioinformatics can be explained with the help of immunology, which covers various aspects like humans, pathogens, high-throughput methods, epidemiology, and the prevention of disease. Advances in omics technology, immune modulations, and complex system analysis have significantly affected vaccinology and immunology [64]. Rapidly evolving technology has increased the knowledge about the human immune system and pathogens. The development of a vaccine is a complicated task and has to fulfill specific criteria. The vaccine should be safe, effective, economical to produce, and easy to store. Continuously increasing data and knowledge about pathogens have posed new challenges in the process of vaccine development, such as strict safety and regulatory requirements.

AI-based methods enable the easy execution of complicated tasks based on logical reasoning and the simulation of tedious wet lab-based experiments. In a single run, thousands of vaccine targets can be easily screened from the data of viral proteins. However, predictions made from this virtual screening need to be validated with selected wet lab-based experiments. These automated experiments are executed on the basis of specific analytical tools for prediction and simulation. After appropriate analysis, either the vaccine component or formulation can be produced. These methods are based on block entropy that assists in the identification, assessment, and quantification of conserved regions within a large number of sequences. These conserved regions are further assessed for immunological properties via in silico peptide binding assays, and vaccine targets are identified (Tables 2.2 and 2.3).

TABLE 2.2

The Areas of Bioinformatics Where AI Can Be Applied

	Applications					
	Data and Knowledge Management	Information Extraction	Gene and Noncoding RNA Prediction	Protein Structure Prediction	Evolutionary Studies	Drug Discovery
[18,19]	Yes	No	No	No	Yes	No
[20]	Yes	Yes	No	No	Yes	No
[21]	No	No	Yes	Yes	Yes	No
[22]	No	No	No	Yes	Yes	No
[42,43]	Yes	Yes	No	No	Yes	No
[50,51]	No	Yes	No	No	No	Yes

In this table a comparative analysis is shown of existing research articles and application areas.

2.5 CONCLUSION AND FUTURE PROSPECTS

Bioinformatics, in combination with AI techniques, has the potential to solve complex biological problems and enables the machine to think like a human. In this paper, we have discussed the significant contribution of AI techniques to addressing the challenges associated with various bioinformatics applications. AI has enabled the development of more sophisticated and efficient algorithms to execute multiple tasks. Various software tools have been employed to parse genome sequences and biological data. AI technology involving machine learning, knowledge discovery, and logical reasoning is continuously evolving. AI technology has significantly influenced the development of efficient tools required for the execution of different bioinformatics tasks. The combination of AI and bioinformatics has enabled the rapid and precise analysis of biological data, and the evolution and accurate prediction of 3D protein structures. AI-based bioinformatics has shown a revolutionary impact in the area of vaccine development and has assisted in speeding up the process. In the future, it can further boost the speed of biological research and can impact human health significantly.

The futures of bioinformatics and artificial intelligence are profoundly interconnected. Applications of artificial intelligence in bioinformatics are everlasting and can open new avenues in biological studies. Bioinformatics will play a significant role in the analysis of the massive datasets by exploiting AI-based tools to save time and resources. The amalgamation of AI and bioinformatics will speed up sequence analysis and drug discoveries, especially in medicine and robotic surgery. This combination has the potential to bridge the gap between available protein sequences and known 3D structures. The application of AI will assist in solving the issues associated with the elucidation of the mechanisms of genetic disease. Thus, AI will help in the development of drugs to combat severe genetic diseases and will have a significant impact on human health.

TABLE 2.3

The Major Prediction Model with Their Functions, Advantages, and Limitations

	Function	Proposed Model	Pros	Cons	Input	Reference
1	Predict unknown circRNA-RBP interaction pairs	Employing positive unlabeled learning (P-U learning) model named circRB	1. Effective and accurate prediction model for RBP binding sites 2. Identification avoids learning bias due to elimination of off-target nucleic acid sequence	Does not perform well with unknown data	Various lengths of circRNA fragments	65
2	Drug mechanism of action easily via deep learning	Online tool called genetic profile-activity relationship (GPAR) implementing deep learning	1. Easy prediction of mechanism of drugs' action via deep learning 2. Minimizes the influence of batch effects and noise with a large number of samples 3. More precise classification	1. Performance of GPAR is not equal for all MOA prediction 2. The availability of drug target is also a limitation	Transcriptome data (GSE92742) from the LINCS dataset project	66
3	Drug-induced cell viability prediction from LINCS-L1000	WRFEN–XGBoost algorithm based on random forest and elastic nets	1. Highly effective compared to existing models 2. Highly robust method 3. Identifies the biomarkers to distinguish between drug-sensitive cell lines and drug-resistant cell lines	Requires more cross validation with large sample size	Transcriptomics signatures (LINCS-L1000), the Cancer Therapeutics Response Portal (CTRP), and NCI-60 dataset	67
4	Predicts the response and efficacy of different anticancer drugs to breast cancer	Deep learning model, and an unsupervised variational autoencoder model gene VAE and rectified junction tree variational autoencoder	Efficient and accurate prediction model	Sample size and sample availability should be considered		68

(Continued)

	Function	Proposed Model	Pros	Cons	Input	Reference
5	DNA:RNA triplex forming potential prediction	An integrated program named Triplex Forming Potential Prediction (TriplexFPP)	1. Avoids redundancy 2. Explores functions of lncRNA	1. Small data size 2. Needs to explore and verify more data 3. R-loop forming type fraction lncRNA in collected data may influence the results	lncRNAs verified by bio assays to form triplexes with DNA from the peer-viewed publications	69
6	Protein kinase inhibitor response prediction	Quantitative Structure–Mutation–Activity Relationship Tests (QSMART) model and neural networks framework	1. Predicts protein–protein interactions associated with the JNK apoptotic pathway 2. Identifies association between lung development and axon extension	Sample size and sample availability should be considered	PKI response dataset	70
7	Biomedical named entity recognition with syntactic information	BIOKMNER, a BioNER model for biomedical texts with key-value memory networks (KVMN) to incorporate auto-processed syntactic information	1. Auto-processed syntactic information 2. Outperforms the existing strong baseline method	Cross validation required	BioNER datasets	71
8	Prediction and prioritization of autism-associated long noncoding RNAs	Developed a support vector machine (SVM) model Our models, including logistic regression, support vector machine and random forest	1. ASD risk genes can be accurately predicted 2. Provides information for the functional characterization of the candidate lncRNAs associated with ASD 3. Minimizes model overfitting	Sample collected from normal brain tissue, not autistic brain samples	ASD risk genes collected from the Simons Foundation Autism Research Initiative (SFARI) database	72

(Continued)

TABLE 2.3 (CONTINUED)
The Major Prediction Model with Their Functions, Advantages, and Limitations

	Function	Proposed Model	Pros	Cons	Input	Reference
9	Predicting MiRNA-disease associations	Novel multiple meta-paths fusion graph embedding model	1. Performs better than other prediction methods 2. Identifies miRNA–disease associations	Requires more information in heterogeneous network	Cannot predict miRNAs associated with breast neoplasms and lung neoplasms	73
10	Phenotype prediction-based on gene expression	Deep neural network model built from gene expression data	1. Produces interpretations more consistent with biology compared to other advanced technologies 2. Performs better than classical machine learning methods on large training sets	Additional biological analysis, comparison and validation that are necessary to get a comprehensive picture of the logic behind the neural network predictions	Microarray data	74
11	Predict miRNA-disease associations	Matrix decomposition and collaborative filtering	1. Identifies the missing miRNA–disease association 2. Performs better than previous methods	1. Materials used contains noise and outliers 2. The least square error function is unstable with noises and outliers 3. Need to perform wet lab experiments to verify the predictions	HMDD V2.0 database	75

BIBLIOGRAPHY

1. S. A. Sehgal, A. H. Mirza, R. A. Tahir, and A. Mir, "*Quick Guideline for Computational Drug Design*," 2018; Bentham Science Publishers.
2. D. Gusfield, S. Eddhu, and C. Langley, "Optimal, efficient reconstruction of phylogenetic networks with constrained recombination," *Journal of Bioinformatics and Computational Biology*, Vol. 2(01), pp. 173–213, 2004.
3. S. A. Sehgal, S. Mannan, and S. Ali, "Pharmacoinformatics and molecular docking studies reveal potential novel antidepressants against neurodegenerative disorders by targeting HSPB8," *Drug Design, Development and Therapy*, Vol. 10, pp. 1605, 2016.
4. N. A. Sehgal, A. Khattak, and Mir, "Structural, phylogenetic and docking studies of D-amino acid oxidase activator (DAOA), a candidate schizophrenia gene," *Theoretical Biology and Medical Modelling*, Vol. 10(1), pp. 3, 2013.
5. S. K. Sehgal, R. A. Tahir, Z. Khalid, and M. A. Hammad, "In silico elucidation of potential drug target sites of the Thumb index Fold Protein, Wnt-8b," *Tropical Journal of Pharmaceutical Research*, Vol. 17(3), pp. 491–497, 2018.
6. Y. Dali, S. M. Abbasi, S. A. F. Khan, S. A. Larra, R. Rasool, Q. T. Ain, and T. H. Jafar, "Computational drug design and exploration of potent phytochemicals against cancer through *in silico* approaches," *Biomedical Letters*, Vol. 5(1), pp. 21–26, 2019.
7. N. Coudray, P.S. Ocampo, T. Sakellaropoulos, N. Narula, M. Snuderl, D. Fenyo, A. L. Moreira, N. Razavian, and A. Tsirigos, "Classification and mutation prediction from non–small cell lung cancer histopathology images using deep learning," *Nature Medicine*, Vol. 24(10), pp. 1559–1567, 2018.
8. S. A. Sehgal, S. Mannan, S. Kanwal, I. Naveed, and A. Mir, "Adaptive evolution and elucidating the potential inhibitor against schizophrenia to target DAOA (G72) isoforms," *Drug Design, Development and Therapy*, Vol. 9, pp. 3471, 2015.
9. Howe, M. Costanzo, P. Fey, T. Gojobori, L. Hannick, W. Hide, D. P. Hill, R. Kania, M. Schaeffer, S. St Pierre, S. Twigger, O. White, and S. Y. Rhee, "Big data: The future of biocuration," *Nature*, Vol. 455(7209), pp. 47–50, 2008.
10. Y. Han, S. Gao, K. Muegge, W. Zhang, and B. Zhou, "Advanced applications of RNA sequencing and challenges," *Bioinformatics and Biology Insights*, 9s1:BBI.S28991, 2015.
11. P. Khatri, M. Sirota, and A. J. Butte, "Ten years of pathway analysis: Current approaches and outstanding challenges," *PLOS Computational Biology*, Vol. 8(2), pp. 1–10, 2012.
12. F. Castiglione, F. Pappalardo, M. Bernaschi, and S. Motta, "Optimization of HAART with genetic algorithms and agent-based models of HIV infection," *Bioinformatics*, Vol. 23(24), pp. 3–5, 2007.
13. M. Bessarabova, A. Ishkin, L. JeBailey, T. Nikolskaya, and Y. Nikolsky, "Knowledge-based analysis of proteomics data," *BMC Bioinformatics*, Vol. 13(Suppl 16), pp. S13, 2012.
14. J. W. Lee, J. B. Lee, M. Park, and S. H. Song, "An extensive comparison of recent classification tools applied to microarray data," *Computational Statistics & Data Analysis*, Vol. 48(4), pp. 9–15, 2005.
15. K. H. Yu, A.L. Beam, and Kohane, "Artificial intelligence in healthcare," *Nature Biomedical Engineering*, Vol. 2(12), pp. 719–731, 2018.
16. R. A. Tahir, H. Wu, M. A. Rizwan, T. H. Jafar, S. Saleem, and S. A. Sehgal, "Immuno informatics and molecular docking studies reveal potential epitope-based peptide vaccine against DENV-NS3 protein," *Journal of Theoretical Biology*, Vol. 459, pp. 162–170, 2018.
17. B. Hayes-Roth, B. Buchanan, O. Lichtarge, M. Hewette, R. Altman, J. Brinkley, C. Cornelius, B. Duncan, and O. Jardetzky, "PROTEAN: deriving protein structure from constraints," In Proceedings of 5th Natl. Conf. on Artificial Intelligence, Morgan Kaufman, Los Altos, Calif, pp. 904–909, 1986.

18. C. Bizer, T. Heath, and T. Berners-Lee, "Linked data-the story so far," *International Journal on Semantic Web and Information Systems*, Vol. 5(3), pp. 1–22, 2009.

19. S. Nagi, D. K. Bhattacharyya, and J. K. Kalita, "Complex detection from PPI data using ensemble method," *Network Modelling Analysis in Health Informatics and Bioinformatics*, Vol. 6(1), pp. 1–13, 2017.

20. F. Rinaldi, O. Lithgow, S. Gama-Castro, H. Solano, H., A. Lopez-Fuentes, L. J. Muniz Rascado, C. Ishida-Gutiérrez, C. F. Mendez-Cruz, and J. Collado- Vides, "Strategies towards digital and semi-automated curation in Regulon DB," *Database*, bax012, pp. 1–11, 2017.

21. A. M. Michel, K. R. Choudhury, A. E. Firth, N. T. Ingolia, J. F. Atkins, and P. V. Baranov, "Observation of dually decoded regions of the human genome using ribosome profiling data," *Genome Research*, Vol. 22(11), pp. 2219–2229, 2012.

22. A. Drozdetskiy, C. Cole, J. Procter, and G.J. Barton, "JPred4: A protein secondary structure prediction server," *Nucleic Acids Research*, Vol. 43(W1), pp. W389–W394, 2015.

23. S. Wang, S. Sun, Z. Li, R. Zhang, and J. Xu, "Accurate de novo prediction of protein contact map by ultra-deep learning model," *PLOS Computational Biology*, Vol. 13(1), pp. 1–34, 2017.

24. E. Faraggi, and A. Kloczkowski, *"Accurate Prediction of One-Dimensional Protein Structure Features Using SPINE-X,"* Springer, New York, pp. 45–53, 2017.

25. S. Wang, W. Li, R. Zhang, S. Liu, and J. Xu, "Coin Fold: A web server for protein contact prediction and contact-assisted protein folding," *Nucleic Acids Research*, Vol. 44(W1), pp. W361–W366, 2016.

26. Q. Wuyun, W. Zheng, Z.Peng, and J. Yang, "A large-scale comparative assessment of methods for residue–residue contact prediction," *Briefings in Bioinformatics*, Vol. 19(2), pp. 219–230, 2018.

27. D. T. Jones, T. Singh, T. Kosciolek, and S. Tetchner, "MetaPSICOV: Combining coevolution methods for accurate prediction of contacts and long range hydrogen bonding in proteins," *Bioinformatics*, Vol. 31(7), pp. 999–1006, 2015.

28. M. Torrisi, and G. Pollastri, "Protein Structure Annotations." In: N. A. Shaik, K. R. Hakeem, B. Banaganapalli, R. Elango, editors, *"Essentials of Bioinformatics, Volume I: Understanding Bioinformatics: Genes to Proteins,"* 2019; Cham, Springer International Publishing, pp. 201–34.

29. F. Meng, V. N. Uversky, and L. Kurgan, "Comprehensive review of methods for prediction of intrinsic disorder and its molecular functions," *Cellular and Molecular Life Sciences*, Vol. 74, pp. 3069–3090, 2017.

30. J. Hanson, Y. Yang, K. Paliwal, and Y.Zhou, "Improving protein disorder prediction by deep bidirectional long short-term memory recurrent neural networks," *Bioinformatics*, Vol. 33, pp. 685–92, 2017.

31. Z. Yuan, "Better prediction of protein contact number using a support vector regression analysis of amino acid sequence," *BMC Bioinformatics*, Vol. 6, pp. 248, 2005.

32. M. Steinegger, M. Meier, M. Mirdita, H. Vhringer, S. J. Haunsberger, and J. Sding, "HHsuite3 for fast remote homology detection and deep protein annotation," *BMC Bioinformatics*, Vol. 20, pp. 473, 2019.

33. M. S. Klausen, M. C. Jespersen, H. Nielsen, K. K. Jensen, V. I. Jurtz, C. K. Snderby, M. O. A. Sommer, O. Winther, M. Nielsen, B. Petersen, and P. Marcatili, "NetSurfP-2.0: Improved prediction of protein structural features by integrated deep learning," *Proteins: Structure, Function, and Bioinformatics*, Vol. 87, pp. 520–527, 2019.

34. M. Blaszczyk, M. Jamroz, S. Kmiecik, and A. Kolinski, "CABS-fold: Server for the de novo and consensus-based prediction of protein structure," *Nucleic Acids Research*, Vol. 41(W1), pp. W406–W411, 2013

35. S. Shatabda, M. A. H. Newton, and A. Sattar, "Constraint-based evolutionary local search for protein structures with secondary motifs," *PRICAI: Trends in Artificial Intelligence*, pp. 333–344, 2014.

36. B. Dogan, and T. Olmez, "A novel state space representation for the solution of 2D-HP protein folding problem using reinforcement learning methods," *Applied Soft Computing*, Vol. 26, pp. 213–223, 2015.

37. S. P. Dubey, N. G. Kini, S. Balaji, and M. S. Kumar, "Protein structure prediction on 2D square HP lattice with revised fitness function," In Proceedings of International Conference on Advances in Computing, Communications and Informatics (ICACCI), pp. 1732–1736, 2017.

38. N. Yanev, M. Traykov, P. Milanov, and B. Yurukov, "Protein folding prediction in a cubic lattice in hydrophobic-polar model," *Journal of Computational Biology*, Vol. 24(5), pp. 412–421, 2017.

39. J. Peng, and J. Xu, "Low-homology protein threading," *Bioinformatics*, Vol. 26(12), pp. 294–300, 2010.

40. L. C. Xue, J. A. P. Rodrigues, D. Dobbs, V. Honavar, and A. M. Bonvin, "Template-based protein–protein docking exploiting pairwise interfacial residue restraints," *Briefings in Bioinformatics*, Vol. 18(3), pp. 458–466, 2017.

41. S. Traore, K. E. Roberts, D. Allouche, B. R. Donald, I. Andre, T. Schiex, and S. Barbe, "Fast search algorithms for computational protein design," *Journal of Computational Chemistry*, Vol. 37(12), pp. 1048–1058, 2016.

42. F. Coste, "Learning the Language of Biological Sequences," 2016; Springer, Berlin Heidelberg, Berlin, pp. 215–247.

43. D. Sundfeld, C. Razzolini, G. Teodoro, A. Boukerche, and A. C. M. A. de Melo, "Pa-star: A disk-assisted parallel a-star strategy with locality-sensitive hash for multiple sequence alignment," *Journal of Parallel and Distributed Computing*, Vo. 112(2), pp. 154–165, 2017.

44. M. Omar, R. Salam, R. Abdullah, and N. Rashid, "Multiple sequence alignment using optimization algorithms," *International Journal of Computational Intelligence*, Vol. 1(2), pp. 81–89, 2005.

45. M. R. Aniba, O. Poch, A. Marchler-Bauer, and J. D. Thompson, "Alexsys: A knowledge-based expert system for multiple sequence alignment construction and analysis," *Nucleic Acids Research*, Vol. 38(19), pp. 6338–6349, 2010.

46. J. I. Requeno, and J. M. Colom, "Evaluation of properties over phylogenetic trees using stochastic logics," *BMC Bioinformatics*, Vol. 17(1), pp. 235, 2016.

47. L. Koponen, E. Oikarinen, T. Janhunen, and L. Saila, "Optimizing phylogenetic super-trees using answer set programming," *Theory and Practice of Logic Programming*, Vol. 15(4–5), pp. 604–619, 2015.

48. T. Le, H. Nguyen, E. Pontelli, and T. C. Son, "ASP at Work: An ASP Implementation of PhyloWS. In: Dovier, A. and Costa, V. S., editors, "Technical Communications of the 28th International Conference on Logic Programming (ICLP'12)," volume 17 of Leibniz International Proceedings in Informatics (LIPIcs), Schloss Dagstuhl–LeibnizZentrum fuer Informatik, Dagstuhl, Germany, pp. 359–369, 2012.

49. N. M. Franz, M. Chen, S. Yu, P. Kianmajd, S. Bowers, and B. Ludascher, "Reasoning over taxonomic change: Exploring Alignments for the perelleschus use case," *PLOS ONE*, Vol. 10(2), pp. 1–34, 2015.

50. E. Ford, K. St. John, and W. C. Wheeler, "Towards improving searches for optimal phylogenies," *Systematic Biology*, Vol. 64(1), pp. 56–65, 2015.

51. André C. A. Nascimento, Ricardo B. C. Prudêncio, and Ivan G. Costa, "A multiple kernel learning algorithm for drug-target interaction prediction," *BMC Bioinformatics*, Vol. 17(46), pp. 1–16, 2016.

52. G.Schneider, "Automating drug discovery," *Nature Reviews Drug Discovery*, Vol. 17, pp. 97–113, 2017.

53. Nadine Bakkar, Tina Kovalik, Ileana Lorenzini, Scott Spangler, Alix Lacoste, Kyle Sponaugle, Philip Ferrante, Elenee Argentinis, Rita Sattler, and Robert Bowser, "Artificial intelligence in neurodegenerative disease research: use of IBM Watson to identify additional RNA-binding proteins altered in amyotrophic lateral sclerosis," *Acta Neuropathol*, Vol. 135, pp. 227–247, 2018.

54. M. H. S. Segler, M. Preuss, and M. P. Waller, "Planning chemical syntheses with deep neural networks and symbolic," *AI. Nature*, Vol. 555, pp. 604–610, 2018.

55. T. Rodrigues, M. Werner, J. Roth, E. H. G. da Cruz, M. C. Marques, P. Akkapeddi, S. A. Lobo, A. Koeberle, F. Corzana, E. N. da Silva Júnior, O. Werz, and G. J. L. Bernardes, "Machine intelligence decrypts b-lapachone as an allosteric 5-lipoxygenase inhibitor," *Chemical Science*, Vol. 9, pp. 6899–6903, 2018.

56. A. Lozano-Diez, R. Zazo, D. Toledano, J. Gonzalez-Rodriguez, "An analysis of the influence of deep neural network (DNN) topology in bottleneck feature based language recognition," *PLoS One* 12(8), pp. 1–22, 2017.

57. Jae Yong Ryu, Hyun Uk Kim, and Sang Yup Lee, "Deep learning improves prediction of drug–drug and drug– food interactions," *PNAS*, Vol. 115(18), pp. 4304–4311, 2018.

58. D. Dimova, and J. Bajorath, "Advances in activity cliff research," *Molecular Informatics*, Vol. 35(5), pp. 181–191, 2016.

59. A. B. Raies, and V. B. Bajic, "In silico toxicology: Computational methods for the prediction of chemical toxicity," *Wiley Interdisciplinary Reviews: Computational Molecular Science*, Vol. 6(2), pp. 147–172, 2016.

60. R. Benigni, and C.Bossa, "Structure alerts for carcinogenicity, and the Salmonella assay system: A novel insight through the chemical relational databases technology," *Mutation Research/Reviews in Mutation Research*, Vol. 659(3), pp. 248 – 261, 2008.

61. P. N. Judson, "*Knowledge-Based Approaches for Predicting the Sites and Products of Metabolism*," 2014; Wiley-Blackwell, pp. 293–318.

62. Y. Sakiyama, H. Yuki, T. Moriya, K. Hattori, M. Suzuki, K. Shimada, and T. Honma, "Predicting human liver microsomal stability with machine learning techniques," *Journal of Molecular Graphics and Modelling*, Vol. 26(6), pp. 907–915, 2008.

63. P. Banerjee, V.B. Siramshetty, M.N. Drwal, and R. Preissner, "Computational methods for prediction of in vitro effects of new chemical structures," *Journal of Cheminformatics*, Vol. 8(1), pp. 51, 2016.

64. G. Sun, C. N. Larsen, N. Baumgarth, E. B. Klem, R. H. Scheuermann, "Comprehensive annota- tion of mature peptides and genotypes for Zika virus," *PLoS One*, 12(1), pp. e0170462, 2017.

65. Z. Wang, and X. Lei, "Identifying the sequence specificities of circRNA-binding pro-teins based on a capsule network architecture," *BMC Bioinformatics*, Vol. 22, pp. 19, 2021. https://doi.org/10.1186/s12859-020-03942-3

66. S. Gao, L. Han, D. Luo *et al.*, "Modeling drug mechanism of action with large scale gene-expression profiles using GPAR, an artificial intelligence platform," *BMC Bioinformatics*, Vol. 22, pp. 17, 2021. https://doi.org/10.1186/s12859-020-03915-6

67. J. Lu, M. Chen, Y. Qin, "Drug-induced cell viability prediction from LINCS-L1000 through WRFEN-XGBoost algorithm," *BMC Bioinformatics*, Vol. 22, pp. 13, 2021. https://doi.org/10.1186/s12859-020-03949-w

68. J. Xie, H. Dong, Z. Jing, D. Ren, "Variational autoencoder for anti-cancer drug response prediction," *Bioinformatics*. Preprint at https://arxiv.org/abs/2008.09763?context=cs.L G, 2020.

69. Y. Zhang, Y. Long, C.K. Kwoh, "Deep learning based DNA: RNA triplex forming potential prediction," *BMC Bioinformatics*, Vol. 21, pp. 522, 2020. https://doi.org/10.1 186/s12859-020-03864-0

70. L.C. Huang, W. Yeung, Y. Wang *et al.*, "Quantitative Structure–Mutation–Activity Relationship Tests (QSMART) model for protein kinase inhibitor response prediction," *BMC Bioinformatics*, Vol. 21, pp. 520, 2020. https://doi.org/10.1186/s12859-020 -03842-6

71. Y. Tian, W. Shen, Song, Y. *et al.*, "Improving biomedical named entity recognition with syntactic information," *BMC Bioinformatics*, Vol. 21, pp. 539, 2020. https://doi.org/10.1 186/s12859-020-03834-6

72. J. Wang, and L. Wang, "Prediction and prioritization of autism-associated long noncoding RNAs using gene expression and sequence features," *BMC Bioinformatics*, Vol. 21, pp. 505, 2020. https://doi.org/10.1186/s12859-020-03843-5

73. L. Zhang, B. Liu, Z. Li *et al.*, "Predicting MiRNA-disease associations by multiple meta-paths fusion graph embedding model," *BMC Bioinformatics*, Vol. 21, pp. 470, 2020. https://doi.org/10.1186/s12859-020-03765-2

74. B. Hanczar, F. Zehraoui, T. Issa *et al.*, "Biological interpretation of deep neural network for phenotype prediction based on gene expression," *BMC Bioinformatics* Vol. 21, pp. 501, 2020. https://doi.org/10.1186/s12859-020-03836-4

75. X. Chen, .L. Wang, J. Qu, N. N. Guan, and J. Q. Li, "Predicting miRNA-disease association based on inductive matrix completion," *Bioinformatics*, Vol. 15, pp. 4256–4265, 2018. doi: 10.1093/bioinformatics/bty503

3 AI-Based Natural Language Processing for the Generation of Meaningful Information Electronic Health Record (EHR) Data

Kuldeep Singh Kaswan, Loveleen Gaur,
Jagjit Singh Dhatterwal, and Rajesh Kumar

CONTENTS

DOI: 10.1201/9781003126164-3

3.1   INTRODUCTION

Thanks to the Health Information Technologies for Economic and Clinical Health (HITECH) Act (2009), which offered opportunities for institutions and healthcare practitioners to implement EHR systems for $30 billion, hospitals have moved on EHR data adoption systems over the past ten years [1]. Accordingly, almost 84% of hospital settlements have implemented at least one basic EHR program, a nine-fold rise from 2008, as latest in this survey by the Office of the Regional Supervisor for health informatics (ONC). However, the usage of simple and accredited EHRs by doctors in offices will more than increased from 42% to 87% [3]. Information relating to the patient's activities, including personal records, symptoms, diagnostic testing, medical reports, medications, radiologic photographs, medical records, and much more [1] are contained in EHR databases. In addition, other experiments have found secondary interest in therapeutic computer science applications [4, 5], predominantly for the enhancement of health quality in an organizational way. Patient records in EHR systems, in particular, have been used for tasks like clinical definition extracting [6, 7], pathway analysis of patients [8], disease inference [9, 10], help in medical decision-making [11], and much more (Table 3.1). Until recent years, conventional machine research and computational methods, for example, technical correlation, SVM, and naive Bayes have been used in most of the methods for the analysis of rich EHR data [12]. Deep learning technology has recently been efficient in several areas through the development of deep hierarchical characteristics and powerful data collection of long-term dependencies [13]. The numbers of reports using deep learning technology for clinical IT activities have also increased

TABLE 3.1

Many Current Deep EHR Projects

Project	Deep EHR Task	Ref.
DeepPatient	Multi-outcome prediction	Miotto [14]
Deepr	Hospital re-admission prediction	Nguyen [19]
DeepCare	EHR concept representation	Pham [20]
Doctor AI	Heart failure prediction	Choi [21]
Med2Vec	HER concept representation	Choi [22]
eNRBM	Suicide risk stratification	Tran [23]

[14,18,20] that shows better result than conventional techniques and require less time for preprocessing and function development.

3.2　RELATED WORK

Two main forms occur (systematic study [45] and biblical analysis) in the field of literature review. Bibliometric analyses is more useful for the assessment of broad literature data than for a comprehensive analysis. They have also been commonly implemented in various academic fields [42, 44, 46, 47]. Arici et al. [48] addressed study patterns and bibliometric results on the improved truth in science education, for example, on 62 papers gathered from Web of Science (WoS) utilizing context analysis. On the basis of the findings, they established many often-used terminologies, including science teaching, software learning, and electronic learning, especially mobile learning. Therefore, the most popular descriptive words used were "information," "awareness," "technology education," "experimentation," "action." This was noticed. Therefore, beliefs the education and success of students also were found to be statistically significant for researchers [49].

The Structural Topic Model (STM) was often used to analyze study subjects in a given area of science. For example, Clare and Hickey [50] used the bibliometric analysis and STM approach to examine regional patterns and issues, based on scientific concepts in journal articles on group forestry conducted during the period 1990–2017. This listed 4 key problems of study and 20 themes. The compositions of the yearly themes revealed that the emphasis of the study has changed from international analysis to the area of carbon sequestering. Based on the database of a large group of 16,452 riders, in 2016 and 2017, Lee and Kolodge [51] applied an STM approach to the discovery of 13 topics of study, such as "long-lasting tracking" and "hacking and glitches." We thereby established possible challenges to the use of self-supporting cars that the intellectually disabled experience. it indicated that the government and companies would help in the adoption of automated vehicles in the preparation of public awareness campaigns. A few bibliometric articles are accessible with an emphasis on NLP and clinical trial analysis. Through the NLP standpoint, the NLP biblical interpretation, handheld computing work [54], improved NLP [47], and computational linguistics [55] are available. In comparison, the NLP is a more portable approach.

Clinical study assessment of cites, Tang et al. [56], Chinese conventional medicine-controlled science (CCM), skepticism clinical review report, and controlled randomized tests in radiology [60], all discussed about the prediction of diseases. For example, Quang et al. [57] routinely synthesized findings or outcomes of clinical trials related to Tai Chi for healthcare with a significant number of scientific studies contributing to the efficacy and effects of Tai Chi mind-body workouts. On the basis of regulated clinical research papers from four major Chinese online repositories of TCM cancer diagnosis, Li et al. [58] evaluate the consistency of the literature and to review the facts in the reviews. Rosas [61] also used a framework for the study of the appearance, output, and effect of research papers through the NIHC/NII clinical trials network to determine their existence, success, and effect. Specifically, the papers were assessed in three ways: (1) measurement by means of journal impact tests for the impacts of the documents; (2) evaluation of papers using bibliometric indicators; and (3) examination by contrast with related journals of the results of publications. Sundaram *et al.* [59] used bibliometrics to routinely analyze the progress of sepsis clinical trials. The analysis was more often quoted by scholars from the United States and reported by respected newspapers. In order to examine its attributes and characteristics, Hong *et al.* [60] carried out bibliometric research on radiological RCTs. Based on their study, radiological RCTs were found to have significantly increased their quantity and efficiency, while techniques remained inadequate in consistency. The features and efficiency of chemotherapy RCTs are evaluated by Kim *et al.* [62]. Cramer *et al.* also introduced bibliometrics to provide a detailed review of the features of specific random yoga studies, with a growing number of RCTs exploring the therapeutic utility of yoga interventions [63]. Cramer *et al.* [64] tried to show that the yoga RCTs appeared more successful because they were produced in India. Yoga RCTs carried out in India are 25 times more likely, relative to those carried out elsewhere, to draw optimistic assumptions. In this respect, Indian studies should be treated cautiously and the option of yoga for patients in other regions should be tested. Bayram *et al.* [65] have listed and described the most widely read articles in emergency rooms on ultrasound tests.

Many experiments have raised arguments on topics relating to clinical science that is improved with NLP. Demner-Fushman and Elhadad [11] addressed the comprehensive study of papers on the implementation of NLP in clinical and customer textual information, by taking into consideration all implementations or analytical studies that exploit texts to promote healthcare and meet the needs of clients. Their study has demonstrated that while clinical NLP continues to have functional usefulness for the processing of large living health information, the use of NLP in medical applications has become essential in everyday life. Zeng et al.[66] examined statistical phenotypic reliance analysis utilizing novel NLP approaches for better EHRs from NLP. Different interpretations were gathered by their analysis. For example, the most popular strategies were keywords queries, rule-based strategies, and supervised machine-based NLP techniques. Third, excellently defined search criteria and rules-based structures are typically very specific. Fourth, controlled procedure classifications were able to achieve greater precision and were more comfortable. In fact, advancement in the area of machine learning, for example, profound learning, has become popular (Table 3.2).

3.3 ARTIFICIAL INTELLIGENCE

Today's AI techniques are named "narrow" or "poor" AIs for diagnostic techniques. For instance, photographs of skin lesions may be categorized into diagnostic groups, or a genomic diagnosis may be given by the integration of genomic and phenotypic evidence. They are not artificial intelligence systems and are not adequately robust to tackle certain clinical diagnosis activities. Bridge learning approaches can also be used to modify a completely qualified AI algorithm to conduct similarly necessary activities. It is better shown by the developments in machine vision and neural

TABLE 3.2

Current Review of NLP-Enhanced Clinical Trial Research and Its Related Topics

Reviewer(s) and Year	No. of Articles	Methods	Period	Research Goals
Sheikhalishahi et al. (2019) [18]	43	Systematic review	2007–2018	To thoroughly review the development and uptake of the application of NLP in dealing with unstructured clinical notes data regarding chronic diseases.
Dreisbach et al. (2019) [19]	21	Systematic review	2003–2017	To summarize studies concerning the application of NLP and text mining to extract and process symptom information contained in electronic patient-authored texts.
Demner-Fushman and Elhadad (2016) [11]	Less than 100	Systematic review	Till 2016	To summarize studies concerning the applications of NLP in processing clinical and consumer-produced texts.
Gonzalez-Hernandez et al. (2017) [20]	149	Systematic review	Till 2017	To summarize studies concerning the applications of NLP for processing health-associated information contained in EHRs and social media posts.
Névéol and Zweigenbaum (2015) [21]	154	Systematic review	2014	To review and select the best papers concerning clinical NLP.
Luo et al. (2017) [22]	48	Systematic review	2000–2015	To summarize the state-of-the-art in methodology advances of NLP for EHR-based pharmacovigilance.
Koleck et al. (2019) [23]	27	Systematic review	1999–2017	To summarize studies concerning the applications of NLP for processing and analyzing symptom information contained in EHR narratives.
Kreimeyer et al. (2017) [5]	86	Systematic review	2006–2016	To summarize currently-in-use NLP solutions for clinical texts that could encode free textual data into standardized clinical terminologies and to capture common data elements to fill specified templates.
Velupillai et al. (2015) [24]	Less than 100	Systematic review	2008–2014	To summarize the latest advances in clinical NLP, particularly focusing on semantic analysis and its supporting subtasks.

networks which have been equipped for general visual recognition tasks. Therefore, in the first phase in developing clinical diagnostic AI algorithms, the standard medical function is normally mapped towards a broader problem class. Here we discuss these question groups and quickly emphasize how these approaches interact with genomics.

- **AI Devices**

    AI systems fall primarily into two broad categories. In Machine learning data is organized such as visual, genomic, and EP data. ML techniques aim to aggregate patient characteristics in hospital situations [17]. In the second group, strategies for natural language processing (NLP) derive details from large amounts of data like clinical reports or case records to complement and enhance organized health records. For improved displays, the diagram in Figure 3.1 illustrates the path from clinical data production to treatment decisions, by NLP data optimization, and ML predictive analytics, through the use of machine-readable structured information. They state that the path chart begins and finishes with clinical research. We must be inspired by health issues and adapt to support medical practice, as effective as AI technologies can be.

    It is used to improve the scientific techniques for the power of AI programs for anatomy, biology, EP, or EMR. Long *et al.*, examined ocular pictures to diagnose cataract disorder in congestion [24] and diagnosed diabetic retinopathy from retinal fundus images.
- **Computer Vision in AI**

    Computer vision is a scientific area focused on image and/or video creation, rendering, and interpretation. Computer vision algorithms absorb and synthesize (or "convolute") large-dimension picture data to generate

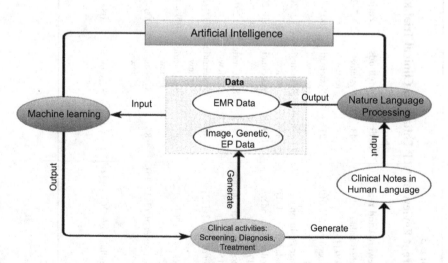

FIGURE 3.1 Framework path from clinical data production to treatment decisions.

quantitative or conceptual depictions of image-incorporated concepts. This method is meant to emulate the manner in which people recognize trends and derive relevant elements from objects. The key stages of image recognition are the collection, data preparation, retrieval and analysis of images, and picture pattern recognition. Deep learning algorithms like Convolutional Neural Network (CNN) are needed to execute vision tasks in a machine. A standard CNN template with smaller matrices called kernel modules or filtering is an activation function. That filter encodes a pattern of image pixels which it "senses" in the source images. A huge number of filters representing different image intensity patterns are distributed throughout the frame to construct two-dimensional activation maps of each sensor. This will then progressively predict the existence of more complicated features using the sequence of features found around the picture (Figure 3.2). Monitoring, image processing, and automated vehicles, are the most significant image

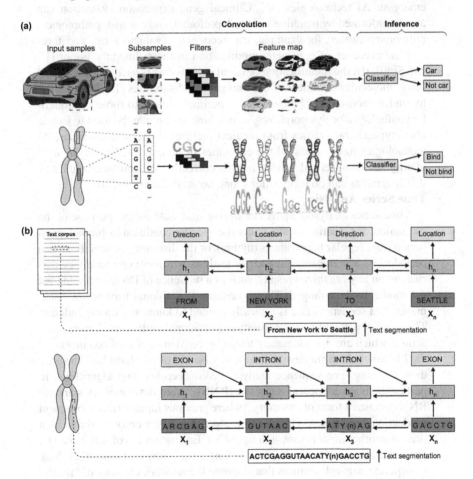

FIGURE 3.2 Monitoring clinical image processing and automated vehicles.

analysis technologies. For diagnostic techniques, implementations of mobile device-vision for biomedical scans (for example, MRI or positron emission tomography images), as well as pictures of pathologies (e.g., histopathermic slides), were regulated by the first AI systems to be approved by the United States Food and Medicine Administration (FDA). Automatic blood circulation quantitative analysis by cardiovascular MRI is the first use of image analysis [7], the calculation of the echocardiogram ejection fraction [8], the identification and volumetric quantitative analysis of chest radiography in radiographs [7], mammography identification, and quantitative analysis of breast densities [9], stroke diagnosis, brain bleeding, and other C-substances. FDA-cleared program for diagnostic uses in pathology.

The pathologists' productivity anticipation is used to greatly improve the entire slides imaging [14] and provide innovative solutions to the automatic diagnosis of dermatological circumstances [15] as well as various other emerging AI technologies [16]. Clinical gene expression evaluation can also be informed by machine vision. A profound study of histopathological pulmonary cancer, for example, can recognize, evaluate form, and forecast the presence of epigenetic modification in the pulmonary tumor [17, 18]. Similarly, the identification of facial expressions is used to classify and direct molecular treatment of unusual genetic disabilities [19, 20]. In order to render decisions for biochemical experiments close to those of a qualified pathologist or dysmorphologist, machine vision may be used to isolate phenotype characteristics from medical pictures. In certain instances, AI technologies have surpassed the capacities of human intelligence by estimating sex from retinal fundus images correctly, for example, a function that specialists can do little better spontaneous devaluation [21].

- **Time Series Analysis**

  Time series analysis requires analyzing time data for the purpose of the estimation of potential events, under the specific condition of physiological period (e.g., regular heart rate vs rhythm), or the discovery of deviations in a period of measurements. Time series analysis will more commonly be carried out on any organized results, such as a sequence of DNA organized but not momentarily arranged. The algorithm of traditional time series bookmarks data sequences and is typically needed to know stream dependency. The increased identification capacity of variational and/or multiphase interactions which are not adequately tested by conventional methods including the Markov-hidden frameworks is the key benefit of AI algorithms in series data scanning. For sequence analysis tasks, deep-learning algorithms, in particular recurrent neural networks (RNNs), were developed. A standard RNN contains a form of "memory," where previous inputs affect subsequent production in a series. The learning algorithm of an entry to the hidden state of another input is shown in figure 3.3. Enhancements of this principle, applied in advanced networks, such as long-term storage networks (LSTMs), incorporate network features that improve the network capacity of "recall" long-term dependencies in entering data. In the event of determining the

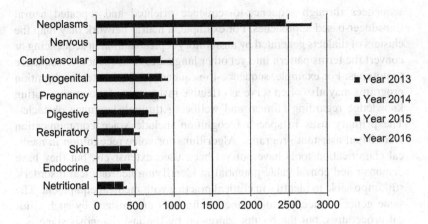

FIGURE 3.3 Searching diseases through classical ML.

distinct state or background that creates the sequential data sequence, CNN models are also extended to time series results. Time series research has significant uses in stock market modeling, atmospheric patterns, and natural phenomena and in theory every potential value case. Time series data AI algorithms for diagnostic techniques may be used in medical instruments that deliver ongoing performance signals, where electrocardiograms are especially involved. AI algorithms are used to recognize and diagnose rhythms [22], especially atrial fibrillation [23] and adipose dysfunction [25] as well as blood chemistry related to heart rhythm abnordics, may be identified and categorized by electrocardiogram [25]. AI time series algorithms tend to be extremely successful when applying genomic sequence data to identify operational DNA sequences suggesting gene salinization [26, 27] and large-scale genetic regions [28] and epigenetic regulation [29].

- **Automatic Speech Recognition**

    A collection of methods and techniques to understand the spoken language involves automated natural language processing. Expression-recognition algorithms capture raw sound vibrations and analyze them such that simple language components such as speed, rhythm, timbre, intensity, and more nuanced features of expression, like tongues, terms, and phrases, can be identified [30]. Further innovative voice recognition algorithms may be used for amazing characteristics such as shifts in behavior or emotion [31, 32] from auditory information. Regardless of the time complexities of expression, standard algorithms use different versions to reinsert value from the spoken expression. Such measures involve in mining audio from distinctive sound units (such as phonemes), linking them to language units (such as words), and transferring them to specific language components (such as phrases) such that definitions are derived. Such functions are now performed in one prototype with synchronized performance [33, 34] with current revelations in the sense of AI algorithms that deal with temporal sequences through sequence-to-sequence oriented and repeated neural transducer-based approaches. For example, a neural network may map the clusters of dialects generated by an auditory representation in sequencing or convert the terms pattern into yet other languages in sequence-to-sequence prototypes. For example, sequence-to-sequence and other text-recognition programs may also often serve as effective instruments for communicating knowledge regarding clinical and wellbeing through language obstacles. The primary uses in speech recognition include voice communication and virtual assistant programs. Algorithms for voice recognition in medical classification tools have not yet been used extensively, but they have demonstrated considerable potential in identifying neurological disorders, still impossible to identify in clinical practice with medical equipment. The same generic voice/reconnaissance techniques are utilized by such clinical procedures, but the results addressed by the last diagnosis stage is a malady phenotype usually correlated with speech (ton, speed, tone, etc.) features and not generally language material. Speech identification of illness with apparent speech impact, including persistent pharyngitis [35] and of disorders of less evident speech effect, including Alzheimer's disease [3], Parkinson's disease [36], manic depression [36], post-traumatic stress disorder [38], and also coronary heart disease [39], has been effectively implemented. Voice recognition can identify possible genetic disturbances, including videos, and. To order to ease the usage of EHRs, language detection should be used as a guide to automated interpretation, to support physicians and patients, and to facilitate an examination of natural language (NLP) [40, 41].

3.4   MACHINE LEARNING OVERVIEW

The path to perfected and unregulated thinking can be separated into two main groups. The guided learning strategies include the projection of y = f(x) from inputs x to y outputs. For starters, correlation and identification using algorithms such as multiple linear regression and support vector machine (SVM) require supervised learning activities. In comparison, the purpose of unattended computer instruction is to think about the spread of x itself. Clustering and intensity calculation are forms of unsupervised training activities. Input processing is a crucial concern with all forms of machine learning systems. Series of attributes identified as characteristics are retrieved for use as a reference for machine learning techniques for each piece of data. These apps are developed utilizing domain information of mainstream machine learning.

3.4.1   APPROACHES TO MACHINE LEARNING

Templates (guidelines) defined often completely lack sweeping generalization consistency and it is often a time-consuming process to manage them properly and revise.

Machine learning methods have generated strenuous concerns for several bioinformatics natural language processing tasks, with greatly increased access to digital bioinformatics tools and powerful computational technology, which can be primarily categorized into five classifications:

- Classification in machine learning: delegate predefined names to documentation.
- Ranking in machine learning: preferably order objects.
- Regression in machine learning: use performance as a projection for the actual value.
- Structured prediction in machine learning: pattern and optimization to classify persons or other semitone structures.
- Clustering in machine learning: explore the fundamental framework of discrete classes with unstructured data.

Of the previously listed activities, various clinical science code programs (medicines, disorders, doses, etc.) may be drafted. EHR analysis may be carried out using formal forecast models; EHRs are examples of detection processes for predicting harmful events. The aim of machine learning is to permit properly predicting parameter values from the respective contexts in the defined true - experimental (attributes or features). In recent years, various methods of learning are used. They may be classified as generative models and biased models in terms of their computational strategies. The generative method structures the common distribution of probabilities through dependent and independent variables such as the naive Bayes, the Bayesian network, the secret Markov model, and the random field Markov.

- **Generative Model**

The generative model is a complete probabilistic framework for all factors which can modulate value systems for any parameter estimates. It can be developed as a conditional probability distribution for classification, using Bayes' theorem. When the illustrated data is small, the generative model is useful for utilizing a vast volume of raw data for better results. Through representing the data, the generative process reduces the uncertainty of the parameter estimate at the cost of potentially adding model bias.

Naive Bayes classifier in generative model: the naive Bayes classification is based on the Bayesian theorem [33] which is a relatively basic generative probabilistic method that can be applied in measuring the likelihood within each applicant class name, provided that each function is categorized separately. It needs just a limited amount of training data for the quicker estimation of parameters, but the strict independence principle for actual implementations is frequently broken, which contributes to a significant distortion.

Bayesian network in generative model: a Bayesian network [34], also a confidence network, is a probabilistic graphical system in which edges are a sequence of probability distributions linked to each of these parameters by a directed acyclic graph (DAG). This technique does not require the assertion of freedom as in naive Bayes. It gives better coverage in real-life implementations and therefore flexibility in the estimation methods. This model the interdependence of variables to have effective handling of lost values and is commonly used in explanatory reasoning applications, such as help in clinical decisions [35] and data processing on gene expression [36].

Hidden Markov model in generative model: the hidden model of Markov (HMM) [37] is a deterministic modeling approach to the Markov method, where the function goes through many unmeasured stage series, generating a succession of hypotheses.

HMM is commonly applied in pattern-labeling (e.g., gene-protein-recognition [38] and biosequence-recognition [39]) activities, for instance. While in certain cases this kind of predictive approach has functioned incredibly well, there are drawbacks. One essential restriction is the requirement that consecutive results are autonomous because the conceptual dependence in the measurement chain cannot be taken into account. Another weakness is the Markov principle itself, namely that the present state just relies on the previous state automatically, which is similarly unsuitable for other issues.

Markov random field (MRF) is the statistical approach to the probability function of a probability distribution that correlates from each node in the network. It is also called a "Markov network" or an undirected graph model [40]. Among these variables there are Markov structures, that provide chart factorization with statistical probabilities.

In view of computing dependence ratios between variables, MRF is close to the Bayesian network. The Bayesian network is a directed graphical design which depicts likelihood representations, that can also be factored into properties of time series that are ideal to catch cause-related connections between variables, while the MRA is a non-directed graphical structure, which has no directivity at each boundary that links a set of edges.

- **Discriminative Model**

The discriminatory model is built, in contrast with the generative model, to include the goal-dependent variables specifically, based on the measuring variables, explicitly calculating the inputs to the corresponding mapping (afterward) and ignoring the corresponding information representation. There are independent factors, sometimes, when adequate annotated knowledge is available, a discriminatory framework offers reliable generalization efficiency. Nonetheless, robust approaches for modeling prior knowledge, form, complexity, etc., are still absent. Moreover, the connections among parameters are not as clear or evident as in the generative model.

Decision tree in discriminative model: a decision tree (DT) [43] is a conceptual binary tree design which demonstrates how well the values of a response variable can be projected using a collection of dependent variables (attributes). A break between a variety of alternatives is reflected in each branch node depending on a particular attribute and each leaf node reflects an option. The incorporation of a decision tree is a top-down mechanism to reduce the quality of knowledge by projecting the knowledge to less performance, thus trying to sacrifice consistency and convenience.

Logistic regression in discriminative model: in 1920, Peral and Reed [46] first invented the supply chain equation, and gradient boosting is a standardized graphical method used to measure the likelihood of an incident occurring by inserting data in a logit equation at the highest probability. The Bayes model is a biased alternative, since it describes the same category of explanatory variables that are conditional. Thanks to its robustness, versatility, and capacity to manage non-linear consequences it was commonly used for diagnosis and monitoring in medicine [47, 48]. Nonetheless, it typically requires more data than traditional regression to produce reliable and relevant outcomes.

Support Vector Machine (SVM) in discriminative model: SVM are often discrete structures which, depending not only on analytical or structured danger minimal rules, are equipped to distinguish pieces of information (instances), i.e., they identify artifacts in groups, but often develop hyperplanes in a region of large dimensions with the greatest margin in various groups. Multiple versions are eventually traced into the same environment and marked as such depending on the side of the hyperplane.

K-nearest neighbor in discriminative model: the K-nearest neighbor (k-NN) rule [54] is a form of instance-based learning, which delays generalized statement outside school. The aim is to attach a rating or rank to the new observation based on the analysis of the nearest marked training set which has been weighted (for regression) or supported (for classification).

Artificial neural network in discriminative model: artificial neural networks [59] are a numerical theory of human cognitive skill which aims for the simulation of biologically neural network architectures and key components. Artificial neurons (processing units) can be linked in the artificial neural systems through multidirectional signal channels in various layers in an ANN model. Even neurons are usually linked in two successive layers.

Conditional random fields (CRFs) in discriminative model: CRFs are a probability model for the naming and segmentation of organized data such as strings, graphs, and lattice structures. The aim is to establish, for a single series of observations, a linear combination of the likelihood over label sequences rather than a combined approximation of both mark and observer series subsets whose representatives both mimic and vary from representatives of a separate subset. It is very important for the clustering algorithm to determine how the matching (or differences) between objects can be defined. Mahalanobis, Euclidsky, Minkowski, and Jeffries–Matusita are representations of Euclidean distance. There are three important groups of clustering approaches: clustering of partitions [70], clustering of hierarchies [71], and hybrid models [72].

- **Unsupervised Clustering**

  The above-described learning techniques are primarily for controlled learning, needing defined data for model testing. Clustering is an unregulated learning process that explores the underlying framework or trend in an unlabeled data set.

- Classical ML

    ML develops data-mining algorithms to derive application functions. Patient "traits" and often diagnostic findings provide inputs into ML algorithms (Figure 3.3).

Baseline details such as age, race, previous illness, etc., and condition-related data such as medical images, hereditary characteristics, EP profiling, physical examination outcomes, clinical signs, treatment, etc., are usually used in a patient's attributes. To compare to these characteristics, the clinical study also gathers patients' medical results. Those provide clinical signs, life rates for the individual, and measures of measurable illness, such as tumor sizes. To correct things, we apply the jth of X_{ij}'s patient and the product of Y_i's concern. ML algorithms may be subdivided into two main categories, unattended learning and supervised learning, based on how the tests are to be implemented. Unchecked training is famous for the extraction of features while supervised learning is ideal for predictive modeling by creating certain interactions among individual patients (as input) and outcome (as output). More recently, semi-controlled learning, a combination of non-controlled learning and guided learning, was suggested, which is ideal for situations where there are no outcomes on such topics. Figure 3.4 describes all three kinds of instruction. Two big, unattended training techniques are clustering and principal component analysis. Groups of topics with identical features are grouped in cluster centers without utilizing the data. The clustering algorithms provide the clinicians with group labeling by optimizing and reducing patient similarities within the cluster. K-means clustering, hierarchical clustering, and Gaussian mixture clustering are also common clustering algorithms. PCA is primarily used for the reductions of dimension, especially if the phenotype is reported in a huge variety of dimensions, such as genes for an analysis of the genome.

PCA forecasts the statistics on a few primary (PC) paths. Occasionally PCA can be used again to eliminate redundant size.

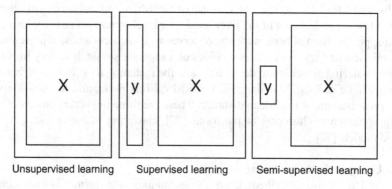

FIGURE 3.4   Unsupervised, supervised, and semi-supervised learning.

In comparison, guided research takes into account the results of the topics along with their features and undergoes specific development phases in order to determine the possible effects of the samples that are similar to the findings on average. The performance formulas typically differ with the tests of concern, e.g., a possibility of a certain health occurrence (Figure 3.5), the predicted value of clinical symptoms, or the anticipated period of life will be the result.

Apparently, supervised instruction produces many validated clinical outcomes in contrast to unregulated learning; for this, AI systems in health coverage use supervised learning. The technical aspects include linear regression, logistic regression, naive Bayes, decision tree, nearest neighbor, random forests, classification analyses, SVMs, and neural networks. (Note that the use of unregulated training as parts of the pre-evaluation phase will minimize the dimension of the data or define subsets, which in effect allows the monitoring supervised learning method to be most effective.) Figure 3.5 illustrates the prevalence of the different supervised learning methods in the field of medicine, showing that SVM and the neural network are by far the most common. As shown in Figure 3.6, we will include more information on the processes of SVM and neural networks along with applications in the fields of cancer, neurological, and cardiac disorders. SVM also applicable in the cases of the three main categories of data (image, genetic, and EP).

- **Support Vector Machine**

Specifically, SVM is used to assign people into the following groups: $Y_i = -1$ or 1 show that the participant is in category 1 or 2, respectively. The basic premise is that the topics can be divided into two classes by a decision limit based on X_{ij} characteristics, which can be written in the following terms (the approach can be generalized to situations of more than two groups):

$$a_i = \sum_{j=1}^{p} w_j X_{ij} + b$$

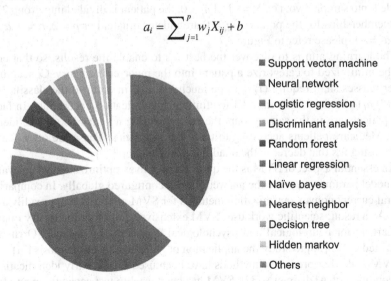

- Support vector machine
- Neural network
- Logistic regression
- Discriminant analysis
- Random forest
- Linear regression
- Naïve bayes
- Nearest neighbor
- Decision tree
- Hidden markov
- Others

FIGURE 3.5 Structure of ML algorithm use in healthcare.

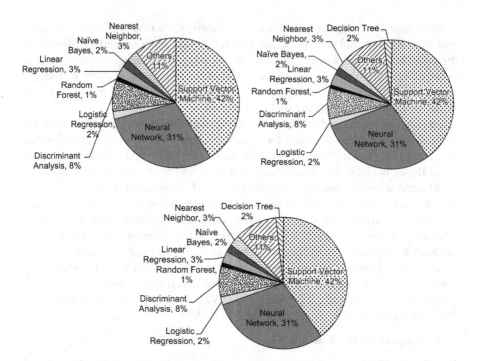

FIGURE 3.6 ML algorithm use in image (upper), genetic (middle), and electrophysiological (bottom) data.

Where w_j is weighting the jth characteristic in order to demonstrate its potential impact on the result. Then, the rule of judgment follows that, if $a_i > 0$, the patient is divided into sample two, i.e., $Y_i = -1$; if $a_i < 0$, the patient is divided into group 2, i.e., the memberships for the points of $a_i = 0$ are indeterminate. For $p = 2$, $b = 0$, $a_1 = 1$, and $a_2 = -1$, please refer to Figure 3.7.

The learning aim is to discover the best w_js to enable the results, so that errors can be minimized to categorize a patient into the incorrect category. Conceptually, better masses shall require (1) a_i to be much like a Y_i in order to the classification, and (2) a_i to be far from 0 in order to eliminate classification uncertainty. In fact, if new patients are in the same group, the corresponding w_js may be used to identify these Medicare patients according to their characteristics. These are often feasible by choosing w_js that decrease the quadratic loss function [29].

An essential aspect of SVM is the question of convex optimization when evaluating model parameters, since the answer often is configured globally. In comparison, several current convex optimization methods for SVM application are readily available. As a result, scientific work uses SVM extensively. To order to classify imaging biomarkers for neurological and psychological disorders, for instance, Orru et al. submitted SVM [30] to study the application of SVM to cancer diagnosis [31].

SVM and other predictive methods have been used for the early identification of the condition of Alzheimer's [32]. SVM has been used to test capacity on an offline man/machine system monitoring upper prothesis [22].

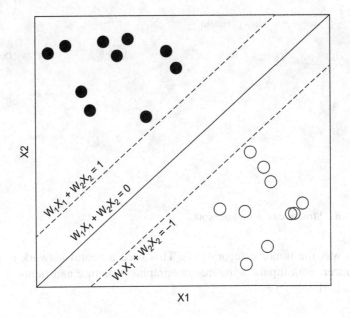

FIGURE 3.7 Structure of SVM.

- **Neural Network**

A neural network may be viewed as an extension of normal distribution to establish linear equation connections amongst inputs and behavioral outcomes. Within the neural network several hidden layer configurations from predefined functionalities reflect the correlations between the outcomes and the entry variables. The target is to approximate the weights using inputs and outcomes data in order to reduce the error percentage of the result. In the following instances, we define the process.

Throughout their study, the input parameters Xi1, allied to the neural network throughout strokes diagnosis [33]. The $p = 16$ signs of stroke include arm or leg paresthesia, severe fatigue, hearing, balance problems, etc. The consequence Y_i is binary: $Y_i = 1/0$ shows that the patient is not shock. The performance parameter of concern is the possibility of incident, i.e.,

$$a_i = h\left\{\sum_{k=1}^{D} w_{2i} f_k \left(\sum_{l=1}^{P} w_{1l} X_{il} + w_{10}\right) + w_{20}\right\}$$

The w_{10} and w_{20} ensure the result from above equation only if both of X_{ij} and f_k are zero; the w_{1l} and w_{2l}s are the weights that display the actual value on the product of the associated multiplicand; the prescribed functions of f_ks and hare illustrate how the weighed combinations impact the infection risk overall. The f_ks here prove that their results are fairly small. Figure 3.8 displays a stylized diagram.

Khan *et al.*'s prediction of cancer was carried out using specific methods whereby the PCs calculated to be inputs in 6,567 genes were used, the product

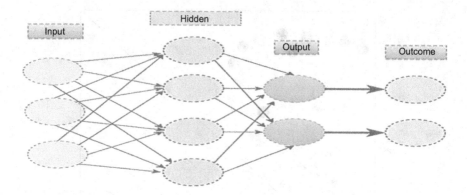

FIGURE 3.8   Structure of neural network.

of which was the tumor category [34]. This used a neural network to forecast breast cancer, with inputs being mammographic structure and tumor predictive outcomes [35].

3.5   DEEP LEARNING OVERVIEW

Profound research requires a broad variety of strategies. This offers a quick summary of the fast-growing profound learning strategies in this portion. We highlight a central equation for every particular architecture that explains the simple operating system. See the exhaustive research of Goodfellow *et al.* [28] for a more thorough description. Classification is the most important concept of fundamental thinking. Traditionally, entry functionality to a machine learning algorithm must be created using raw data to evaluate the trends of prior interest, depending on practitioners' experience and field awareness. The learning strategy to develop, study, pick, and test suitable characteristics can be lengthy and complex and is therefore known as a "black art" [29], which involves imagination, experimentation, and sometimes chance. By comparison, deep learning algorithms acquire the best patterns in images themselves without human input, such that the latent user connections that otherwise would not be identified or concealed may be found automatically.

Through deep learning, complicated data description is sometimes represented as complexes of smaller, simpler representations. For example, it is possible to identify an individual in a picture and identify representations of pixel, curves, and angle margins and facial structure of corners and contours of the edges [28]. This idea of unregulated hierarchy with rising sophistication is a recurring topic in deep learning. Much in the artificial neural network (ANN) draws on the fundamental understanding of algorithms and architectures. ANNs consist of a collection of linked nodes (neurons) grouped in layers as shown in Figure 3.9. Neurons which are not found in layers of input or output shall be recognized as secret modules. Each secret package stores a set of *w*-weights updated to the specification.

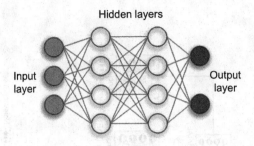

FIGURE 3.9 One input, output layers, and two hidden layers in neural network.

ANN vectors are enhanced by reducing uncertainties like the continuous random variable probability in the equation:

$$E(\theta,D) = -\sum_{i=0}^{D}\left[\log P(Y = y_i \mid x_i,\theta)\right] + \lambda \|\theta\| p$$

In this equation, the first term decreases the amount of register failure over the whole trainee data set D; in the second cycle, the p-norm of the learning model parameters is minimized, regulated with a tunable Ś parameter. This second concept, called regularization, is used to stop the overpowering of a paradigm and to improve the capacity to generalize to new, unseen instances. Usually the loss function is optimized by back propagation, which minimizes losses from the final layer backward via the network [28]. The loss function is optimized. There are several techniques available to work with deep learning techniques in a range of languages.

In the continuation of this segment, we discuss a variety of common forms of deep learning models for deep EHR applications focused on ANN optimization and architecture. Start with supervised techniques (multi-layer perceptron, neural convolutions, and recurring neural networks) and end with unattended architectures (such as auto codecs and Boltzmann-retreated machines). Figure 3.10 provides a hierarchical analysis of certain deep learning frameworks for the study of EHR data and select works carried out in this sample.

3.5.1 MULTI-LAYER PERCEPTRON (MLP)

A multi-layer perceptron is an ANN form composed of several secret layers in which each layer I neuronal is fully linked to the others in layer $i + 1$. In general, these networks are restricted to certain unknown layers and, unlike repeated or undirected structures, information flows in just one direction. Trying to extend a single layer ANN, each concealed element calculates a modified total of the activation function outputs, accompanied by a non-linear activation of the calculated total, as seen in the equation below. Here d is the quantity of items in the preceding layer, x_j is the product of the jth node in the preceding layer, and w_{ij} and b_{ij} are weight and prejudice

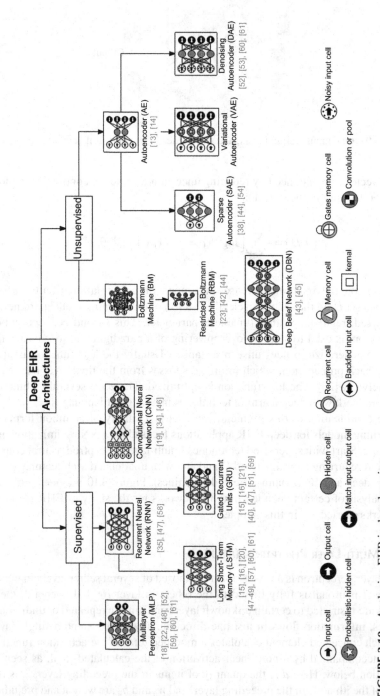

FIGURE 3.10 Analyzing EHR in deep learning.

terms for each x_j. Learning functions are sigmoid or tanh, but current networks often employ functions like linear corrected group (ReLU) [28].

$$h_i = \sigma\left(\sum_{j=1}^{d} x_j w_{ij} + b_{ij}\right)$$

The network can learn to combine input X with output Y after the optimization of hidden layer weight during exercise. Once hidden layers are introduced, the input data is supposed to be gradually symbolic due to the nonlinearity detections of each hidden layer. While MLP is among the easiest, certain implementations almost always integrate neurons in their final layers completely linked with it.

3.5.2 CONVOLUTIONAL NEURAL NETWORKS (CNN)

In recent times, convolutional neural networks (CNNs), particularly in the congregation of image analysis, have become a very common tool. CNNs require special original data interconnection. For example, more relevant attributes may be derived by displaying the object in a set of the local instead of presenting a 50×50 picture as 2500 pixel resolution. One or higher-dimensional time series can often be used as a set of specific signaling. The equilibrium is seen in the following equation where x is the input signal and w are the measurement feature or coevolutionary layer.

$$C_{1d} = \sum_{a=-\infty}^{\infty} x(a) w(t-a)$$

In the equation given, x is a two-dimensional layout (e.g., an image) and K a kernel; this is often seen as a two-dimensional convolution. A kernel or buffer slide a set of weights to remove the function maps through the entire data.

$$C_{2d} = \sum_{m}\sum_{n} X(m,n) K(i-m, j-n)$$

Scarce connections exist in the CNNs, which usually results in very few parameters because the detectors are narrower than the data. Convolution often facilitates the exchange of parameters as each filter is added to the entire file. In a CNN, a coevolutionary layer has a number of coevolutionary filters that are described above and receive the very same input from an earlier layer. In order to combine derived features during such convolutions, a pooling or subsampling layer is typically used. Figure 3.11 shows an example of the two-convolutional-layer CNN design, each of them accompanied by a pooling line.

3.5.3 RECURRENT NEURAL NETWORKS

Recurrent neural networks (RNNs) are the rational option, when the input layer has a simple spatial pattern (such as pixels in an image), and where data is organized

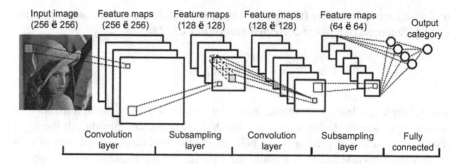

FIGURE 3.11   Structure of two CNN layers with pooling and subsampling layers.

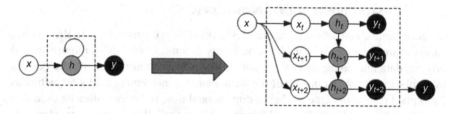

FIGURE 3.12   Framework of symbolic representation of RNN.

sequentially (such as time-series data or natural language) RNNs are the best move. Although a one-dimensional series can be fed into a CNN, the resultant features derived are shallow [28], because the characteristic projections are only influenced by strongly clustered interactions with many neighbors. RNNs are meant to resolve these spatial long-range dependencies. RNNs function by updating the secret HT state, not only dependent on the triggering of the current input at t, but also on the secret HT − 1 state in the previous state. The switch of xt − 1, HT − 2, etc., has been revised (Figure 3.12). That is because the final secret state incorporates knowledge from all the previous components, after storing a whole chain.

Long short-term memory (LSTM) and gated recurring unit (GRU) versions, also called gated RNNs, are common RNN variations. Although typical RNNs consist of integrated concealed cells, a cell that comprises an inner recurrence loop, and a gate structure that governs the flow of knowledge is substituted by each unit of a gated RNN. In modeling long-term sequential dependence, the gated RNNs have shown benefits between other advantages [28].

3.5.4   Auto Encoders (AE)

The autoencoder is one of the fundamental learning frameworks that demonstrates the concept of unmonitored representations research. They were already made famous as an initial tool for pretraining supervised profound learning models, especially, if classification techniques are limited but still remain useful in totally unattended tasks, such as the discovery of phenotypes. Autoencoders are meant to encrypt the data to a smaller dimension z. The encoded image is then decrypted by reshaping an estimated image + of the input x. The encryption and restoration method as illustrated in the two calculations below, with an a single considered to be more accurate because encryption and decoding weights and the reconstruction error is reduced.

$$z = \sigma(Wx + b)$$

$$\bar{x} = \sigma(W'z + b')$$

Once AE has been equipped, the network provides a single input with the most inner secret layer activations that are an encrypted representation of the input. AEs convert the interface data into the form that only saves the much more significant extracted dimensions. These therefore are close to the traditional techniques for dimensional decrease (PCA) and single values (SVD), but have significant advantages for complex issues due to highly nonlinear transforms through the classification algorithms of the hidden layer. A biased method called stacked can be used to build and prepare deep AE networks (Figure 3.13). It is different Variants of EIs were added, including denoising autoencoders (DAE) [32], sparse autoencoding (SAE), and variational autoencoding (VAE) [28].

3.5.5 Restricted Boltzmann Machine (RBM)

The restricted Boltzmann system (RBM) is just another unattended deep-education architectural style to acquire data input representation (Figure 3.14). The aim of RBMs is close to the that of autoencoders, but RBMs also consider the propagation chances of the data input from a stochastic viewpoint. Thus, RBMs are often seen as generational models that attempt to model the process underlying the production of the data. The canonical RBM [28] is an energy modeler with visible binary devices $\sim v$ and hidden devices $\sim h$, with the energy function indicated below in the equation.

$$E(v, h) = -b^T v - c^T h - W v^T h$$

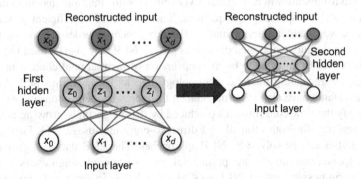

FIGURE 3.13 Structure of stacked AE with hidden layers.

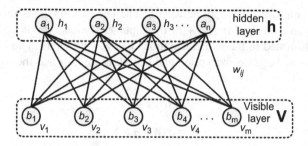

FIGURE 3.14   Structure of RBM.

All units are linked absolutely in a typical Boltzmann system, while in the RBM there are two obvious units or two secret units that are not attached. RBM training is normally carried out through stochastic optimization including sampling with Gibbs. It results in a final *h* form, which is supposed to support the initial input data. A deep belief network (DBN) may be hierarchically stacked for supervised learning algorithms.

3.6   NATURAL LANGUAGE PROCESSING (NLP)

NLP is the real human language digital abstraction of content. Such algorithms use a text input or possibly automated voice recognition performance and result in a valuable content transformation. This transition may involve the translation of languages, description of records, and reviews or compilations of higher-level texts. Typical NLP alga-ray involves a syntactical analysis that includes scans of written text in a wide range of ways to retrieve computational descriptions of language (for example, through sentence breach, marking parts of language, and optimizing of influenced word forms). Based on the goal performance, from sequence-to-sequence networks and other RNN language processing variants [42] to CNNs, a broad range of neural network architectures have been built to provide better text-to-text interpretations [43].

The diversity of synonyms, words, and interconnected concepts that may convey a single sense is a significant problem discussed by natural language processing. For clinical environments where regulated vocabularies are multiple and in continuous flow, this issue is especially conspicuous. Therefore, NLP is efficient at standardizing these words automatically and synthesizing them in order to forecast present and potential diagnosis and medical events [4, 44]. NLP may also be used to promote exposure to health knowledge by the conversion of instructional content into other languages or translating medical terminology into its lay meanings [45]. AI-based virtual assistants have also been introduced to improve genetics consultants' capacity to satisfy the time commitments produced by the increasingly growing amount of genetic research for both clinical and direct-to-consumer usage [46]. Furthermore, EHR analytics can be solved by NLP approaches. The EHR data is highly dimensionally, sparse, incomplete, and prejudicial, and create confusing factors. For starters, following hospitalization, NLP was added to EHRs in order to estimate patient mortality. Throughout this procedure, EHR details are translated into a set of medical incidents which have been conveyed to an RNN and have been qualified to recognize trends in medical symptoms, diagnosis, demography, treatment, and other incidents forecasting short-term prognosis or hospital readmission in patients [4]. Likewise, estimates of symptom incidence and treatment efficacy may be produced when paired with other scientific evidence [47]. In tandem with genomic evidence, NLP approaches have been used for the estimation and the conduct of knowledgeable genetic diagnosis of unusual disease diagnoses, contributing to predictive genetic assessments that are reliable and close to those of human analysts [48, 49].

- **Foundations of Biomedical Natural Language Processing**

NLP is an area of study that helps computers with the right information to interpret natural language content, in the end rendering communication among computers and humans possible with various forms of natural languages. From biomedical articles in genetics, medicine, and chemistry, the subcategory of bioengineering NLP is defined. In each of these fields, there is a wide variation in vocabulary as expressed in their specific literature, rules, etc. Moreover, the same kinds of bioengineering messages, as previously discussed, could differ tremendously due to variations in expression and certain organization-specific variance.

Theory and solutions to linguistic and machine learning lay solid foundations from which effective NLP systems can be established in a number of real-world scenarios. While some methods and templates for biomedical NLP implementations are mentioned below, they can all be applied to clinical research electronic records (EHRs).

- **Sublanguage Approach in Biomedical Natural Language Processing**

Grishman [22] describes a sublanguage as a specific natural language used to represent a restricted topic in which a community of subjects' expertise is usually working. The word sublanguage, which is the algebra as the fundamental formalism, was one of the first to be mentioned by Zellig Harris [20]. The language is described as a subsystem of the language which is closed for any or all communication processes.

The principle of secondary vocabulary offered a framework for natural language processing in specific situations, such as therapeutic accounts. Numerous natural language processing frameworks establish the minimal domain syntax and somaticizing of sublanguage functions. An EHR is limited to concerns of patient treatment because, unlike scientific research, it is unlikely to contain gene descriptions or issues with the cell line. In contrast to the ordinary words, sublanguages have several unique features, culminating in complex terminology, structured structures, and specific persons and relationship issues.

- **Vocabulary Level in Biomedical Natural Language Processing**

A language that is very distinct from the norm is a professional vocabulary. For nonbiological texts, for example, "cell side" is rarely alluded to. Specifically, the advancement of science and technical advances in the biomedical sector has contributed to the identification, through studying sublanguages in the respective corpus, of new biological artifacts, roles, and occurrences.

- **Syntax Level in Biomedical Natural Language Processing**

A sublanguage is not just an arbitrary collection of words, it is not just special in grammar and vocabulary structures. For nursing, for example, telegraphic expressions such as "enhanced patient" are phonetic because of procedures to enable objects and auxiliaries to be sign for understand the patient behavior. There are also in the sublanguage several speech structures consisting of predicate words and organized parameters like "<antibody> <appeared in> <tissue>"; in predicated phrases, <appeared> and in the semantine-related terminology as like"<antibody>" and "<tissue>."

- **Discourse and Semantics Level**

A sublanguage may often provide specific forms to perceive and coordinate larger speech units alongside the variations between vocabulary and syntax stages. For examples, "secondary" implies that an explanatory relationship is different to its use in normal language. The hierarchical structure also consists of the context of existing illnesses, entry pharmaceutical items, social record, physical inspection, etc.

These sublanguage properties allow for the usage of interpretation and retrieval techniques which cannot be used in the production of newspaper headlines or novels' sentences. Sublanguage research may also combine subject awareness with current structures. For example, a biotechnology retrieval method can be established by annotating medical journals solely on context from a collection of words that are known to concern researchers. The regulated medical vocabulary can be extracted by means of a sublanguage analysis centered on terms which frequently combine with similar terms.

- **A Brief "Description" of Existing Clinical Natural Language Processing Systems**

EHRs rely on the heterogeneous essence of clinical documents, including organized explanations and accounts, to collect accurate patient details. There have been huge efforts in the last two decades to improve natural language processing biomedical structures for the extraction of clinical narratives. Many of the methods were discussed in two forms. Regulatory methods rely on the usage of sublanguage modeling and trends, while machine learning strategies explore various useful functions and appropriate algorithms. A subject information tool is typically used by both strategies.

- **Rule-Based Approach in Clinical Natural Language Processing Systems**

EHRs rely on the heterogeneous essence of clinical documents, including organized explanations and accounts, to collect accurate patient details. In the last 2 decades, enormous efforts have been made to enhance biomedical frameworks for the extraction of clinical narratives in natural language processing. Much of the methods were discussed in two forms. Regulatory methods rely on the usage of sublanguage modeling and trends, while machine learning strategies explore different useful functions and suitable algorithms. A subject information tool is typically used by both strategies. A grammar consists of precisely described semantic constructs, their meanings, and the underlying target constructs. The analyser is supported by a MedLEE approach, which is meant to decrease uncertainty in a domain language by integrating the template with semantic approaches owing to its basic meaning. The MERKI regulatory framework for extracting drug names and associated characteristics from organized and descriptive medical transcripts was developed [80]. Recently [81] an automated drug extractor program (MEDEx) was developed to take stock of malfunctions using semantic rules and a map reader to obtain promising findings on prescription extraction and the associated fields such as power, pathways, size, shape, dose, and length. This documentation was described by a simple semantic model for medication type of drug observations, to which drug text was connected.

- **Learning-Based Approach in Clinical Natural Language Processing Systems**

SymText is a learning-oriented method focused on NLPs, which involves a transfer syntactic translator, built on improved process networks and translation terminology, and a Bayesian network-based semantic model [34], used to derive results from chest radiography [83] in a variety of applications. In both the genomic information and the clinical reports, Agarwal and Yu created two biomedical NLP systems called NegScope [84] and HedgeScope [85], which were able to distinguish negative and hedge markers as well as their scopes. Both systems have been focused on the CRF [64], the publicly accessible BioScope [86] corpus. A learning model has been developed. Lancet [87] is a supervised machine-learning program which automatedly extracts from clinical discharge summaries drug events comprised of medicine names and information related to their prescription usage (dosage, style, frequency, length, and reason). In order to decide the drug names and fields belonging to a single medicine case, Lancet employs the CRFs [64] model, and the AdaBoost model, which includes the decision stump algorithm [88]. Lancet produced the highest precision amongst top-ten systems during its third i2b2 sharing assignment for natural language processing for medical information retrieval dilemmas. Cao *et al.* [89] developed a clinical question response program named AskHERMES, a computer framework that automatedly examines the clinical queries, and retrieves and mines vast literature records and clinical notes concerning particular issues, in order to assist healthcare professionals in reacting to queries that occur during their encounters with patients rapidly and efficiently. For questions analysis [90], the problem classifier and search terms recognition assisted vector machines (SVMs) were implemented in the form of a learning algorithm [49] and CRFs model [64].

The framework is built to require healthcare professionals in medical settings to effectively search for knowledge. Liu *et al.* [91] built a clinical environment language recognition program, Clinical ASR, for more efficient access to knowledge such as a clinical response system, to offer a voice interface to clinical NLP applications. Language modification (LM) on the SRI decipher method [92] is discussed by Clinical ASR on the basis of clinical problems.

3.7 ELECTRONIC HEALTH RECORD SYSTEMS (EHR)

In hospital and outpatient facilities, the use of EHR systems has increased considerably [2, 3]. The use of EHRs in hospitals and clinics is important to strengthen healthcare delivery by eliminating delays, continuing to increase productivity, and enhancing the coordination of care, while providing researchers with a rich source of data [25]. Without clinical notes, basic EHR systems with clinical notes and comprehensive systems are typically grouped into basic EHRs [2]. Simplistic EHR programs may offer a variety of knowledge about the patient background, complications, and usage of drugs though missing sophisticated features. Since EHR has been mainly developed for management activities within the hospital, there are many labeling systems and regulated terminology to document specific patient knowledge and incidents. Many instances of this are medical codes, such as ICD, treatment codes such as Current Procedural Terminology (CPT, for example), test tests such as conceptual identifications for analysis (LOINC), and drug codes such as RxNorm (Table 3.3). Such codes include hierarchical systems of part-mappings, such as the United Medical Language System (UMLS) and the Systematic Nomenclature of Medicine-Clinical Terms (SNOMED CT). In view of the broad range of systems, a continuous field of study is the harmonization and interpretation of data

TABLE 3.3

Classification Schema Laboratory Tests

Schema	Number of Codes	Examples
ICD-10	68,000	- J9600: acute respiratory failure (diagnosis)
		- I509: heart failure
		- I5020: systolic heart failure
CPT	9,641	- 72146: MRI thoracic spine (procedures)
		- 67810: eyelid skin biopsy
		- 19301: partial mastectomy
LOINC	80,868	- 4024-6: salicylate, serum (laboratory)
		- 56478-1: ethanol, blood
		- 3414-0: buprenorphine screen
RxNorm	116,075	- 161: acetaminophen (medications)
		- 7052: morphine
		- 1819: buprenorphine

from different terminologies and organizations. Throughout this chapter, many broad EHR structures suggest types of clinical code representation that allow patient research and implementations to cross-cross more easily. EHR systems store different forms of patient records, including clinical, diagnosis, physical tests, and sensor details.

In Table 3.3 categorical principles, such as race or codes from standardized terminology, such as the diagnosis or CPT procedures of the ICD-10 (formerly ICD-9), and available-text natural language, such as improvement reports or abstracts of discharge. In addition, such data forms may be chronologically arranged to shape the base of time series extracted, for example perioperative vital signals or multimodal medical history. While the recent related papers [24, 26, 27] include much biomedical data including medicines or genomics, this study focuses on these five forms of data present in most current EHR systems.

3.8 DEEP LEARNING-BASED EHR

In this segment, we address the current state-of-the-art therapeutic developments as a consequence of recent advancements in EHR education. Table 3.4 offers an

TABLE 3.4
Working on EHR Deep Learning

Task	Subtasks		Input Data	Models	References
Information extraction	(1)	Single concept extraction	Clinical notes	LSTM, Bi-LSTM, GRU, CNN	[15, 16, 34, 35, 36, 37]
	(2)	Temporal event extraction		RNN + Word Embedding	
	(3)	Relation extraction		AE Custom Word	
	(4)	Abbreviation expansion		Embedding	
Representation learning	(1)	Concept representation	Medical codes	RBM, Skip-gram, AE, LSTM	[23, 36, 14, 18–23, 36, 38–40]
	(2)	Patient representation		RBM, Skip-gram, GRU, CNN, AE	
Outcome prediction	(1)	Static prediction	Mixed	AE, LSTM, RBM, DBN LSTM	[14, 18, 23, 41–43, 19–21, 38, 44–48]
	(2)	Temporal prediction			
Phenotyping	(1)	New phenotype discovery	Mixed	AE, LSTM, RBM, DBN LSTM	[14, 18, 23, 41–43, 19–21, 38, 44–48]
	(2)	Improving existing definitions			
De-identification	Clinical text de-identification		Clinical notes	Bi-LSTM, RNN + Word Embedding	[52, 53]

overview of recent profound EHR learning initiatives and their goals, where activities and tasks are described. Some of the programs and the tests this segment is focused on are private EHR data sets from reputable scientific organizations; however, the Medical Knowledge Mart for Intensive Care (MIMIC), publicly accessible database, and the online clinical report data sets from i2b2 (informatics) are included in many studies used in this study.

3.8.1 EHR Information Extraction in Deep Learning

Medical reporting is complicated in comparison with standardized parts of EHR data generally used for invoicing administrative operations, and principally by healthcare providers for instructive papers. Much clinical documentation like admittance reports, departure summary, and transfer orders are associated with patient meeting. Data from clinical records are particularly difficult to collect due to their unstructured nature. Traditionally, these techniques required a lot of manual programming and ontological mapping, and these tactics were therefore constrained. Some recent studies therefore focused on the collection of clinical facts from in-depth knowledge about healthcare. The principal responsibilities include: (1) removal of the concept, (2) temporary exclusion, (3) exclusion from connection and (4) expansion of the symbol (Figure 3.15).

3.8.1.1 Concept of Single Extraction

A standardized academic word, for example diseases and medicines, or processes, is the most crucial function for the free text in clinics. Many recent projects have employed standard NLP approaches to attempt to do so with different degrees of

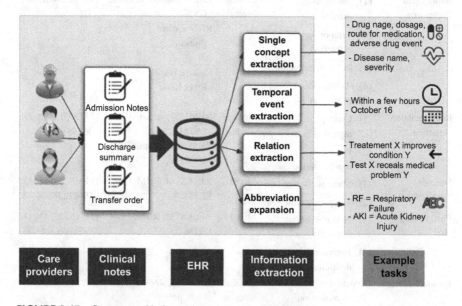

FIGURE 3.15 Structure of information extraction (IE) in EHR.

successful, but the subtle intricacy of the clinical observations allows space for further advancement. Studies in [15, 16] deal with the issue of mining as a series marking procedure to assign each phrase to one of the nine diagnostics labels. We have separated tags in categories of medications and diseases, each label containing tags such as the name of the product, the dose, the treatment route, the presence of dangerous drugs, symptoms, and illnesses. We are working on various deep RNN designs including LSTM and GRU, BI-LSTM and many LSTM variants with standard CRFs. Their findings are linked with reference CRFs which have always been recognized as the advanced technique for removal of medicinal components. They found that all RNN variants are over wide-ranging bounds above CRF baseline techniques, in particularly in the identification of finer characteristics, such as number, strength and disease intensity. This sophisticated information is very helpful for medical information and cannot simply be accessed from the coding step of accounting.

3.8.1.2 Extraction of Temporal Event

This subtask addresses the much more complicated issue of time allocation for Fries [35] established a method to retrieve medical incidents and their associated periods from clinical reports using a computer-trained two major medical companies. For organized interactions and forecasts, Fries used Stanford's Deep Dive application [55]. The SemEval 2016 shared existing challenges and required little hand development.

3.8.1.3 Relation of Extraction

While the abstract time is associated with incidents in health and their duration, complete removal of relationships involves organized interactions between free text medical principles such as X therapy enhances/worsens/causes Y or X studies, which reveals medical issues [36].

3.8.1.4 Expansion of Abbreviation

Over 7,000 distinct scientific abbreviations have been identified in clinical tests [37] which need the addition of formal extraction definitions before they can be described. There can be ten potential theories for any abbreviation, thereby rendering abbreviation extension a daunting job. Liu *et al.* [37] use word integration methods to tackle the issue. Through creating a word2vec based on therapeutic text from intensive care units (ICU), Wikipedia, and medicinal science journals, newspapers, and authors, they build unique terms embedment. While word integration models themselves are not profound models, they are important preconditions for NLP profound learning tasks. In contrast to 82.3% accuracy, this streamlined methodology far exceeded the simple methods of expansion by abbreviation.

3.9 REPRESENTATION OF LEARNING IN EHR

A large range of revealed medical codes representing all facets of patient experiences are included in today's EHR programs. These codes were originally used for internal logistical and accounting functions, but provide valuable details on secondary computer technologies. Presently, a hand-made scheme is used to map organized scientific terms in which each term is given a distinct tag by its accompanying ontology. The underlying semblances among various definition forms and coding schemes cannot be quantified by such rigid hierarchical connections. New methods of deep learning have been used to transform distinct codes into vector spaces for more comprehensive research and analytical activities. We define deep EHRs (e.g., I509) as real-life arbitrary dimension vectors to reflect distinct health codes in this segment. We have defined them as the ICD-10 codes to display administrative medical codes. Both initiatives have been mostly unattended and are based on natural connections and fragments of vector space codes. Detailed methods to patient representations using such codes may be addressed in the next paragraph since doctors can be applied as a well-ordered medical codes. Implementations for patient representations usually improve a supervised learning process by enhancing the description of the information (e.g., patients) to the deep learning network. The mechanisms can often benefit patients.

(1) **Concept of representation**: several recent studies have implemented profound unmonitored strategies of classification problems to extract EHR term vectors capturing latent correlations and functional clustering among primary therapy. We refer to this field as the representation of EHR definition, and its main aim is to extract sparse medically coded vector representations such that related concepts are in close proximity in the smaller dimension region. If such vectors have come into existence, codes with heterogeneous origins (for example prescriptions, medications), such as t-SNE [18, 19, 23], word including representation with unequal health codes [20], or code similarity heatmaps [40], may be categorized and qualitative evaluated.

Distribution of Medical Codes: because clinical terms are sometimes documented in time stamps, a single experience can be interpreted as a series of covert, sentence-like, and word-list medical codes. Several investigators have used NLP methods to synthesize fragmented safety codes in vector graphics with a set and compact scale. One of these methods is called skip-gram, a pattern popularized by Mikolov *et al.* [54]. Word2vec is an unsupervised ANN structure for the acquisition of vector document representation given that the word recognition is a certain dataset, and is also utilized in various fundamental language learning models as a pre-processing phase. [18, 22] and Y, likewise. [39] All use skip-gram for dispersed data embedding in the sense of clinical codes. During the study[39] the problem is handled by dividing the patient list of codes into smaller pieces, randomized the instance list for each section of the slice, which involves numerous therapeutic codes that are allotted to stamps at the same time.

Encoding latent: apart from NLP-inspired approaches, many common depth representation strategies for learning reflect EHR theories are also used. Tran *et al.* [34] created an updated, restricted RBM using a formal training technique to enhance the understanding of representation. In a similar way, Lv *et al.* [36] used AEs to create idea vectors using terms from healthcare providers' language concepts. We measured the intensity of relationships among various medical principles and noticed that relational models of perceptions extracted from EAs alone advanced conventional linear models significantly, achieving cutting-edge efficiency.

(2) **Representation of patient**: in the literature [20, 22, 23, 38, 39] many separate deep learning approaches have been proposed to achieve distributed representation of clinicians. Most notably technologies for dispersed word representations [54], such as NPP strategies or technologies for dimensionality reduction such as autoencoders, are used as a foundation of these techniques [13]. One method influenced by the NLP [18, 22, 38] is to extract distributed vector representations.

Skip grams and repeated neural network models are used in physician paragraphs, i.e., the structured ICD-9, CPT, LOINC, and National Drug Codes (NDC). Similarly, for forecasting unplanned hospital readmission, deep structure uses a simple word embedded layer as an entrance to a broader CNN architecture [19].

3.10 METHODS OF EVALUATION FOR EHR REPRESENTATION LEARNING

Most experiments involving representational learning test participant opinions on the basis of the auxiliary classification activities, indirectly suggesting the prediction advances. A better presentation of certain scientific terms or patients is referred.

In certain research secondary grouping activities are not included and the emphasis is on specifically assessing the learning outcomes. Because there is no accepted criterion for such activities, the appraisal approaches are varied again. The principle of coherence used by Tran *et al.* is initially used in the simulation of the subject [23]. To order to quantitatively analyze the clusters of clinical codes, scientists have developed two personalized measures known as the Health System Similarity Measure (MCSM) and the Clinical Relation Meter (MRM).

3.10.1 OUTCOME PREDICTION IN EHR REPRESENTATION LEARNING

Many profound EHR initiatives mainly aim to predict treatment experience. Two particular forecasting modes are identified: (1) persistent or specialized prediction (e.g., heart disease) using one-time data and (2) time prediction (i.e., prediction of heart attack within six months or forecast of illness beginning from traditional linear encounter data). The greatest contribution in some cases to linear models to evaluate the consistency of derivatives representation is the depth of representing itself in order to increase its effectiveness (Table 3.5).

(1) **Static outcome prediction in EHR representation learning**: without noticing temporal restrictions, the easiest type of outcome prediction program is forecasting a certain outcome. For example, in order to forecast a cardiac collapse, Choi *et al.* use distributed and multiple ANN and linear models [18]. They thought that the right configuration was a regular MLP with embedded patient curves, with raw-categorical coding for all versions.

(2) **Temporal outcome prediction in EHR representation learning**: many experiments have established deeper learning algorithms to predict the outcomes or the start of a forecast, or to predict a forecast based on time

TABLE 3.5
Prediction Outcome Tasks in Deep EHR

Outcome Type	Outcome	Model
Static	Heart failure	MLP [18]
	Hypertension	CNN [41]
	Infections	RBM [42]
	Osteoporosis	DBN [43]
	Suicide risk stratification	RBM [23]
Temporal	Cardiovascular, pulmonary	CNN [44]
	Diabetes, mental health	LSTM [20]
	Re-admission	TCNN [19]
	Heart failure	GRU [21], [38]
	Renal	RNN [47]
	Postoperative outcomes	LSTM [46]
	Multi-outcome (78 ICD codes)	AE [14]
	Multi-outcome (128 ICD codes)	LSTM [45]

series evidence. Cheng et al have performed CNN study on transient matrix codes per patient to monitor both the development of congestive cardiac insufficiency (CHF) and chronic obstructive pulmonary disease (COPD) [44]. They tested with many CNN temporal approaches including heavy, early, and late merge, and noticed that for all prediction tasks the CNN with heavy-fusion outperformed other CNN variants and linear models.

3.11 THE CASE FOR NLP SYSTEMS AS AN EHR-BASED CLINICAL RESEARCH TOOL

We give a brief summary of NLP to provide clinicians with a basic interpretation of NLP and do not address the concern of NLP in NLP studies. Multiple comments from NLP blogs are a matter of worry. NLP is a field of IT, machine linguistic science that allows for the interaction of machine-language and the bridging of the interactions between medical speech in the natural world and computer networks. NLP, ML, and deep (ML) thinking are all part of AI and NLP. It includes insightful thinking and ML. NLP interacts with people via the machine using natural languages (human, not code languages).

Any activities which use natural speech, such as writing (documents) or voiced (speaking) language, to identify the key fundamentals, can be loosely described as NLP. Based on the clinical review, we relate it to legally safe medical information for clinicians that is generated online by doctors, physicians, and a wide variety of other team leaders in the clinical field. There are two solutions to NLP. One is guided (the machine implements the plan's preset guidelines) and the other is ML (the computer knows human latent rules). Guidelines called [supervised] or no [unsupervised] human instruction).

Though useful clinical records (for example, history of current disease, psychiatric background, summary of physical exam results, analysis of the pulmonary function examination, and radiology or surgical reports) are contained in free EHR documents, it is essential to collect and insert these details into organized data. Many, or all, large studies or methods to population management thus rely heavily on standardised data in order to dissuade physicians and analysers from obtaining complete and comprehensive EHR knowledge. The minimal use of large amounts of data in EHRs has helped to keep the data inaccurately organized.

Clinical science and clinical care documentation (e.g., ICD codes). This task can be solved by NLP. As shown in Figure 3.16, NLP cohort detection techniques take information(s) from EHRs, process information collected and then score patients by rule learner. Such analytical methods for the NLP are complicated since the NLP does not consist of a specific method, but instead a diverse grouping of techniques. This involves diligent preprocessing of texts and other NLP functions to translate a document from its original state to a set of terms to transform phrases and words into principles of interest. The following are a few instances of NLP functions at the low level (text preprocessing): (1) sentence boundary elimination, which is described in the normal way by a time (2) tokenization (bridging a phrase into a token) (e.g., translation of "patients have a background of numerous wheezing episodes" into "patients"), (3) stopping (cutting a word into a root form) ("have," "many").

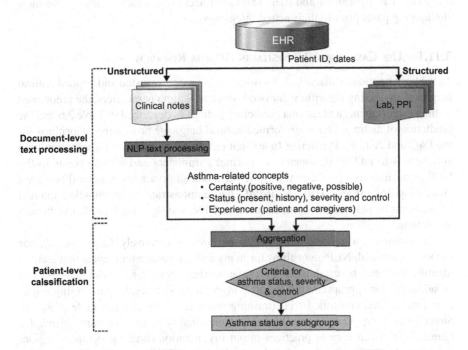

FIGURE 3.16   Flow charts of NLP algorithm extract information from EHR.

Instances of NLP functions at a higher level include:

(1) designated identification agent (identifying particular terms or phrases such as cancer, DNA, drugs) and (2) defining laws on negation. A simple keyword search is not feasible since the simple keyword search does not require specific types of clinician notes to be separated from EHR. The National Library offers a range of well-known information network tools to support NLP activities (such as the UMLS Met thesaurus for documents, synonyms of identification for identified persons, and a term uncertainty text collection). When the retrieving of information and the analysis of EHRs has been completed, the person has to be categorised and aggregated according to established directives (e.g. criteria for asthma) (i.e., part of the production phase of the NLP algorithm).

An expert classification is utilized to construct NLP algorithms as a generic criterion and suggests certain functionalities (e.g., asthma incidents or asthma diagnosis by experts). This move usually requires an annotation tool (such as Anafora). Training a computer or learner typically involves respondents indicating mechanisms until a specialist's interpretation is used to identify the theory of benefit properly (extract, process, and classify them). A developed or qualified NLP algorithm uses a large amount of development and training. Using a self-contained group (e.g., a k-fold cross-validation process (such as train/test split)) the NLP technique is evaluated once the principle is implemented optimally. Since NLP algorithms have been initially developed and tested, the students need to determine furthermore the portability of NLP algorithms and their validation accuracy, which is among the most challenging parts prior to their actual integration.

3.11.1 Use Cases for NLP System in Asthma Research

As we have previously discussed, for instance, we recently tested and applied natural language processing algorithms for two historical asthma parameters, the predefined asthma requirements and asthma prediction indices as described in Table 3.6 and the prediction of asthma. Our team formed natural language processing algorithms for the PAC and Asthma-Predictive Index that employed a manual review section with annotations to add all parameters as normal parameters and asthma status to the NLP algorithms as a prediction. Despite the nuanced principles of natural language processing-PAC from EHRs, NLP-PAC has demonstrated that clinician manual assessments display responsiveness, features, PPV, and significant predictive benefit for asthma.

The estimates were 97%, 95%, 90%, and 98%, respectively. Consequently, our earlier studies with NLP algorithms for many asthma parameters reveal that asthma identification can be established by implementing specific scientific principles in broad scales for appropriate patient NLP algorithms. Further, by performing a specific randomized controlled trial (training sequence 5298 and test code 5297), in Sioux Falls, South Dakota, we illustrated functionality in basic research principles (generalization) in various practices in our organization (broad geographic region, patient community, documentational procedures, and EHR system).

TABLE 3.6
Using Deep Algorithm for Finding Asthma Disease

PAC

Patients were considered to have definite asthma if a physician had made a diagnosis of asthma and/or
if each of the following three conditions was present, and they were considered to have probable
asthma if only the first two conditions were present:

1. History of cough with wheezing, and/or dyspnea, OR history of cough and/or dyspnea plus
   wheezing on examination,
2. Substantial variability in symptoms from time to time or periods of weeks or more when
   symptoms were absent, and
3. Two or more of the following:
   • Sleep disturbance by nocturnal cough and wheeze
   • Nonsmoker (14 years or older)
   • Nasal polyps
   • Blood eosinophilia higher than 300/mL
   • Positive wheal and flare skin test results OR elevated serum IgE
   • History of hay fever or infantile eczema OR cough, dyspnea, and wheezing regularly on
     exposure to an antigen
   • Pulmonary function tests showing one FEV1 or FVC less than 70% predicted and another with
     at least 20% improvement to an FEV1of higher than 70% predicted OR methacholine
     challenge test showing 20% or greater decrease in FEV11
   • Favorable clinical response to bronchodilator

Asthma Predictive Index

Major Criteria	Minor Criteria
(1) Physician diagnosis of asthma for parents	1. Physician diagnosis of allergic rhinitis for patient
(2) Physician diagnosis of eczema for patient	2. Wheezing apart from colds Eosinophilia (≥4%)

In estimating the allergy condition manual map analysis, sensitivities, accuracy, PPV, and negative predictive value (the NPVs of an NLP-PAC algorithm were 92%, 96%, 89%, and 97% respectively) in a separate study.

Mainly, NLP-PAC and NLP-API recognize a community of asthmatic children with atypical clinical and immunological characteristics (e.g., chronic asthma, greater likelihood for asthma exacerbation, respiratory illness, and greater chance of disease).

For the NLP asthma prognosis algorithms (length of cure and recovery and recurrent asthma; remission was calculated by three consecutive years of separation of asthma events) 94%, 98%, and 0.96% were identified as the post-index asthma period, PPV, and F scale. A main function of our asthma and prognostic NLP program is an ability, which is essential for epidemiological studies involving timeliness, to predict the index date for asthma or incidents (e.g., initiation of asthma or recovery or recesses). Such NLP algorithms technologies may have robust reliability or population reduction and that the statistical variability of allergy, asthma, and immunology phenotyping that sometimes obscure true biopsy and dissuade the application of empirical proof thus that clinical pressure.

3.12 IMPLICATIONS OF NLP FOR EHR BASED ON CLINICAL RESEARCH AND CARE

The functional aspects of an EHR comprising fundamental, structured and unorganized clinical details (i.e. laboratory information); and the NLP's multi-modal capacity for collecting, managing and identifying patients that permit significant clinical actions (i.e. accurate columns); (Figure 3.17). Thus, NLP may serve as an effective tool for rendering EHRs a valuable repository of information to meet patients' and researchers' requirements and to encourage essential practices while minimizing the pressure of graph analysis and EHR data extraction (e.g., 70% of EHR physicians have reports of burnout). Clinicians, for example, also have to conform with legislative requirements (e.g., disclosing asthma output to a state authority and the Minnesota population measure). This incorporates accurate measure of asthmatic children as a majority. Artificial asthma intelligence may enable physicians to successfully perform this role and also maintain monitoring of the asthma position provided by NLP, as NLP algorithms provide clinicians verified rationale (i.e., clinical notes supporting the diagnosis of asthma). Another illustration will be that NLP will compile and analyze specific EHR details for patients, allowing automatic chart analysis to exploit free texts in EHRs and then clinicians to enable effective decision-making on asthma treatment by not focusing on manual charting.

Our recent studies indicate that automatic graphic analysis via the natural language processing program decreased patients' clinical decision-making time by around 80%. In the US medical sector, the approach to clinical treatment and study of AI or even other information systems would possibly be a significant guidance. Because artificial intelligence relies effectively on high-quality data input, artificial intelligence will be a foundation for high-quality research and treatment as a data analysis resource for EHRs, minimizing procedural variability in phenotyping, an essential move in allergy and asthma study. A big influence on potential EHRO-driven clinical treatment and science is the ongoing commitment and effort

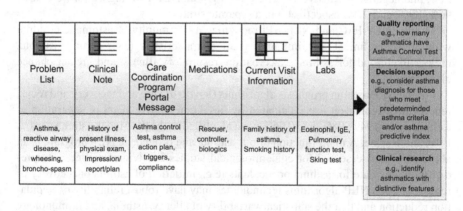

FIGURE 3.17 Framework of finding infected asthma patients using NLP algorithm-based EHR.

in developing the interoperability standards for EHRs guided by the Office of the Regional Advisor for Health IT in compliance with the 21st Century Cures Act.

Recommended system analyzes still present further challenges in the future: problems in terms of data integrity such as incomplete information and imbalances, the security issue (hosting or transitioning EHR data to cloud) algorithms, interoperability requirements and drug-induced burnout, and health-related information and technology operations and maintenance considerations. The workshop on the subject of artificial intelligence in healthcare in 2019, for instance, illuminated trust, clarity, accessibility, accountability and

AI justice as future threats. Although major studies of asthma and allergies are focused largely on standardized data (e.g., ICD code) and are conducted on a routine basis, given the recognized shortcomings as addressed in our literature review conducted, work is currently being conducted on utilizing free text through NLP in EHRs.

This can help us recognize algorithmic distortions arising from the inherent distortions of EHRs, free of structural biases. Given these limitations, EHR-based study, as a source of real-world data, is supplementary to the lately recognized conventional randomized, regulated research proof.

Valuable suggestions have been newly offered for regular adoption and approval workflow for the essential functionalities of a data management framework: (1) the inacceptable usage of black boxes (such as in case of why and how), (2) the loss of time is a cost, (3) sophistication and simplicity are not appropriate for consumers, (4) the utilization of software is necessary, and (5) awareness and information distribution.

3.13 CONCLUSION AND FUTURE DIRECTIONS

The ML algorithms have eclipsed state-of-the-art methods and acquired FDA clarification for a wide spectrum of medical diagnoses, particularly diagnostics based on images. This performance improvement is caused by the accessibility of huge volumes of data for training, for example, a greater number of image processing annotations and large-scale functional genomic data points as well as advancements in the usage of iInformation technology and GPU systems. At present, deep phenotypic knowledge from photos, epidermis, and other medical tools for informing horizontal genome evaluation is the more exciting implementation of AI in medicinal genomics. Fortunately, in a number of clinical biotechnology activities, such as mutation naming, genome classification, and practical effect estimation, deep learning algorithms have also proven extremely promising. More specific AI methods can be common in these fields, in particular where the inference from complicated data (the variants) is an often-repetitive activity for clinical genomics. The developments in CNNs and RNNs, which tend to be especially suitable for the analysis of genome details, have improved several requirements.

However, it has not yet been compellingly shown how effective AI algorithms are as the ultimate help for therapeutic choices for predicting specific complicated psychological phenotypes. The study results needed to achieve this aim might ultimately be included in the increased clinical research efforts for the collection of behavioral Health information, such the British Biobank, [96] and the US Development Program [97]. With the dependency of AI on wide training data systems, it would surely be the flexible array of phenotypic data instead of genetic data that would be the more difficult hurdles for understanding their objectives. DNA-b The genomic data may be created quickly and at level using the basis of sequencing technology. But the data analysis for phenotypic expressions includes many forms of data processing and collects data on collecting sites in a slower, more costly and highly complicated way. In conclusion, the prediction mistake and commitment to system bias are important for the broad adoption of AI technology in every given therapeutic technique.

BIBLIOGRAPHY

1. Torkamani, A, Andersen, KG, Steinhubl, SR, Topol, EJ, "High-definition medicine", *Cell*. 2017;170: 828–43.
2. Esteva, A, Robicquet, A, Ramsundar, B, Kuleshov, V, DePristo, M, Chou K, "A guide to deep learning in healthcare", *Nat Med*. 2019;25:24–9.
3. Fraser, KC, Meltzer, JA, Rudzicz, F, "Linguistic features identify Alzheimer's disease in narrative speech", *J Alzheimers Dis*. 2016;49: 407–22.
4. Rajkomar, A, Oren, E, Chen, K, Dai, AM, Hajaj, N, Liu, PJ, "Scalable and accurate deep learning for electronic health records", *NPJ Digit Med*. 2018;1:18. https://doi.org /10.1038/s41746-018-0029-1.
5. Zou, J, Huss, M, Abid, A, Mohammadi, P, Torkamani, A, Telenti, A, "A primer on deep learning in genomics", *Nat Genet*. 2019;51: 12–8.
6. Eraslan, G, Avsec, Ž, Gagneur, J, Theis, FJ, "Deep learning: new computational modelling techniques for genomics", *Nat Rev Genet*. 2019;20: 389–403.
7. Retson, TA, Besser, AH, Sall, S, Golden, D, Hsiao, A, "Machine learning and deep neural networks in thoracic and cardiovascular imaging", *J Thorac Imaging*. 2019;34: 192–201.
8. Asch, FM *et al.*, "Accuracy and reproducibility of a novel artificial intelligence deep learningbased algorithm for automated calculation of ejection fraction in echocardiography", *J Am Coll Cardiol*. 2019;73(9 Supplement 1):1447. https://doi.org/10.1016/S0 735-1097(19)32053-4.
9. Le, EPV, Wang, Y, Huang, Y, Hickman, S, Gilbert, FJ, "Artificial intelligence in breast imaging", *Clin Radiol*. 2019;74: 357–66.
10. Majumdar, A, Brattain, L, Telfer, B, Farris, C, Scalera, J, "Detecting intracranial hemorrhage with deep learning", *Conf Proc IEEE Eng Med Biol Soc*. 2018 ;2018 :583–7.
11. FDA approves stroke-detecting AI software. *Nat Biotechnol*. 2018; 36:290. https://doi .org/10.1038/nbt0418-290.
12. Gulshan, V, *et al.* "Development and Validation of a Deep Learning Algorithm for Detection of Diabetic Retinopathy in Retinal Fundus Photographs". *JAMA*. 2016;316: 2402–10.
13. Van Der Heijden, AA, Abramoff, MD, Verbraak, F, Van Hecke, MV, Liem, A, Nijpels G, "Validation of Automated Screening for Referable Diabetic Retinopathy with the IDx-DR Device in the Hoorn Diabetes care System", *Acta Ophthalmol*. 2018; 96:63–8.
14. Miotto, R, Li, L, Kidd, BA, and Dudley, JT, "Deep Patient: An Unsupervised Representation to Predict the Future of Patients from the Electronic Health Records", *Scientific Reports*. 2016; 6(April): 26094.

15. Esteva, A *et al.* "Dermatologist-level classification of skin cancer with deep neural networks", *Nature.* 2017; 542:115–8.

16. Niazi, MKK, Parwani, AV, Gurcan, MN, "Digital pathology and artificial intelligence", *Lancet Oncol.* 2019;29: e253–61.

17. Rios Velazquez, E *et al.*, "Somatic Mutations Drive Distinct Imaging Phenotypes in Lung Cancer". *Cancer Res.* 2017;77: 3922–30.

18. Coudray, N *et al.*, "Classification and Mutation Prediction from non-small cell Lung Cancer Histopathology images using Deep Learning", *Nat Med.* 2018;24: 1559–67.

19. Nguyen, P, Tran, T, Wickramasinghe, N, and Venkatesh, S, "Deepr: A Convolutional Net for Medical Records," *arXiv.* 2016, pp. 1–9.

20. Pham, T, Tran, T, Phung, D, and Venkatesh, S, "Deep Care: A Deep Dynamic Memory Model for Predictive Medicine," *arXiv,* 2016, no. i, pp. 1– 27,.

21. Choi, E, Bahadori, MT, and Sun, J, "Doctor AI: Predicting Clinical Events via Recurrent Neural Networks," *arXiv,* 2015, pp. 1–12.

22. Choi, E., Bahadori, MT, Searles, E, Coffey, C, Sun, J, "Multi-layer Representation Learning for Medical Concepts," *arXiv,* 2016, pp. 1–20.

23. Tran, T., Nguyen, TD, Phung, D, Venkatesh, S, "Learning Vector Representation of Medical Objects via EMR-Driven nonnegative Restricted Boltzmann Machines (eNRBM)", *J. Biomed. Inform.* 2015: 54; 96–105.

24. Kreimeyer, K., Foster, M., Pandey, A., Arya, N., Halford, G., Jones, S.F., Forshee, R., Walderhaug, M., Botsis, T., "Natural Language Processing Systems for Capturing and Standardizing Unstructured Clinical Information: A systematic Review", *J. Biomed. Inform.* 2017: *73*; 14–29.

25. Demner-Fushman, D, Elhadad N, "Aspiring to Unintended Consequences of Natural Language Processing: A Review of Recent Developments in Clinical and Consumer-Generated Text Processing", *Yearb. Med. Inform.* 2016: *25*; 224–233.

26. Sheikhalishahi, S, Miotto, R, Dudley, JT, Lavelli, A, Rinaldi, F, Osmani, V, "Natural Language Processing of Clinical notes on Chronic Diseases: Systematic Review", *JMIR Med. Inform.* 2019: 7; e12239.

27. Dreisbach, C, Koleck, TA, Bourne, PE, Bakken, S, "A Systematic Review of Natural Language Processing and Text Mining of Symptoms from Electronic Patient-Authored Text Data", *Int. J. Med. Inform.* 2019: 125; 37–46.

28. Gonzalez-Hernandez, G, Sarker, A, O'Connor, K, Savova, G, "Capturing the Patient's Perspective: A Review of Advances in Natural Language Processing of Health-related text", Yearb. *Med. Inform.* 2017: 26; 214–227.

29. Névéol, A, Zweigenbaum, P, "Clinical Natural Language Processing in 2014: Foundational Methods Supporting Efficient Healthcare", *Yearb. Med. Inform.* 2015; 24: 194–198.

30. Luo, Y, Thompson, WK, Herr, TM, Zeng, Z, Berendsen, MA, Jonnalagadda, SR, Carson, MB, Starren, J, "Natural Language Processing for EHR-Based Pharmacovigilance: A Structured Review", *Drug Saf.* 2017: 40; 1075–1089.

31. Koleck, TA, Dreisbach, C, Bourne, PE, Bakken, S, "Natural Language Processing of Symptoms Documented in Free-Text Narratives of Electronic Health Records: A Systematic Review", *J. Am. Med. Inform. Assoc.* 2019: 26; 364–379.

32. Velupillai, S, Mowery, D, South, BR, Kvist, M, Dalianis, H, "Recent Advances in Clinical Natural Language Processing in Support of Semantic Analysis", *Yearb. Med. Inform.* 2015: 24; pp. 183–193.

33. Hinton, G, *et al.*, "Deep Neural Networks for Acoustic Modeling in Speech Recognition", *IEEE Signal Process Mag.* 2012; 29:82–97.

34. Prabhavalkar, R, Rao, K, Sainath, TN, Li, B, Johnson, L, Jaitly, N, "A Comparison of Sequence-to-Sequence Models for Speech Recognition", In: Proceedings of the Annual

Conference of the International Speech Communication Association, Interspeech; 2017. Available: https://doi.org/10.21437/Interspeech.2017233.

35. Li, Z, Huang, J, Hu, Z, Li, Z, Huang, J, Hu, Z, "Screening and diagnosis of chronic pharyngitis based on deep learning," *Int J Environ Res Public Health*. 2019;16. Available: https://doi.org/10.3390/ijerph16101688.

36. Zhan, A *et al.*, "Using Smartphones and Machine Learning to Quantify Parkinson Disease Severity the Mobile Parkinson Disease Score", *JAMA Neurol*. 2018;75: 876–80.

37. Ringeval, F *et al.*, "AVEC 2019 Workshop and Challenge: State-of-Mind, Detecting depression with AI, and Cross-Cultural Affect Recognition, In: Proceedings of the 9th International on Audio/Visual Emotion Challenge and Workshop. Nice; 2019. pp. 3–12. Available: https:// doi.org/10.1145/3347320.3357688.

38. Marmar, CR *et al.*, "Speech-Based Markers for Posttraumatic Stress Disorder in US Veterans", *Depress Anxiety*. 2019;36: 607–16.

39. Maor, E, Sara, JD, Orbelo, DM, Lerman, LO, Levanon, Y, Lerman, A, "Voice Signal Characteristics are Independently Associated with Coronary Artery Disease", *Mayo Clin Proc*. 2018;93: 840–7.

40. Mohr, DN, Turner, DW, Pond, GR, Kamath, JS, De Vos, CB, Carpenter, PC, "Speech Recognition as A Transcription Aid: A Randomized Comparison with Standard Transcription", *J Am Med Informatics Assoc*. 2003; 10:85–93.

41. Edwards, E *et al.*, "Medical Speech Recognition: Reaching Parity with Humans", In: Karpov, A, Potapova, R, Mporas, I, editors. Speech and Computer. SPECOM 2017, vol. 10458. Cham: Springer. p. 512–24. Available: http://link.springer.com/10.1007/978-3-319-66429-3_51. Accessed 12 Aug 2019.

42. Wu, Y, Schuster, M *et al.*, "Google's Neural Machine Translation System: Bridging the Gap between Human and Machine Translation", arXiv. 2016; arXiv:1609 08144.

43. Collobert, R, Weston, J., "A Unified Architecture for Natural Language Processing: Deep Neural Networks with Multitask Learning", In: ICML '08. Proceedings of the 25th International Conference on Machine learning. Helsinki; 2008, 2008. pp. 160–7.

44. Miotto, R, Li L, Kidd, BA, Dudley JT., "Deep Patient: An Unsupervised Representation to Predict the Future of Patients from The Electronic Health Records", *Sci Rep*. 2016;6:26094. Available: https://doi.org/10.1038/srep26094.

45. Chen, J *et al.*, "A Natural Language Processing System That Links Medical Terms in Electronic Health Record Notes to Lay Definitions: System Development Using Physician Reviews", *J Med Internet Res*. 2018; 20:e26. Available: https://doi.org/10 .2196/ jmir.8669.

46. Kohut, K, Limb, S, Crawford, G., "The Changing Role of The Genetic Counsellor in The Genomics Era", *Curr Genet Med Rep*. 2019;7:75–84.

47. Diller, G-P *et al.*, "Machine Learning Algorithms Estimating Prognosis and Guiding Therapy in Adult Congenital Heart Disease: Data from A Single Tertiary Centre Including 10, 019 Patients", *Eur Heart J*. 2019;40:1069–77.

48. Liang, H *et al.*, "Evaluation and Accurate Diagnoses of Pediatric Diseases Using Artificial Intelligence", *Nat Med*. 2019;25:433–8.

49. Clark, MM *et al.*, "Diagnosis of Genetic Diseases in Seriously Ill Children by Rapid Whole genome Sequencing and Automated Phenotyping and Interpretation", *Sci Transl Med*. 2019;11:eaat6177. Available: https://doi.org/10.1126/scitranslmed.aat6177.

50. Li, H., "Toward Better Understanding of Artifacts in Variant Calling from High Coverage Samples", *Bioinformatics*. 2014;30:2843–51.

51. DePristo, MA *et al.*, "A Framework for Variation Discovery and Genotyping Using Next-Generation DNA Sequencing Data", *Nat Genet*. 2011;43:491–8.

52. Garrison E, Marth G, "Haplotype-Based Variant Detection from Short-Read Sequencing", arXiv. 2012; arXiv:1207 3907.

53. Hwang, S, Kim, E, Lee, I, Marcotte, EM, "Systematic Comparison of Variant Calling Pipelines Using Gold Standard Personal Exome Variants", *Sci Rep.* 2015;5:17875. Available: https://doi.org/10.1038/srep17875.

54. Poplin, R *et al.*, "Universal SNP and Small-indel Variant Caller using Deep Neural Networks", *Nat Biotechnol.* 2018; 36:983–7.

55. Wick, RR, Judd, LM, Holt, KE, "Performance of Neural Network Base calling tools for Oxford nanopore Sequencing", *Genome Biol.* 2019;20:129. Available: https://doi.org/ 10.1186/s13059-019-1727-y.

56. Tang, H, Thomas, PD., "Tools for Predicting the Functional Impact of Nonsynonymous Genetic Variation", *Genetics.* 2016;203: 635–47.

57. Quang, D, Chen, Y, Xie, X, "DANN: A Deep Learning Approach for Annotating the Pathogenicity of Genetic Variants", *Bioinformatics.* 2015; 31: 761–3.

58. Kircher, M, Witten, DM, Jain, P, O'Roak, BJ, Cooper, GM, Shendure, J, "A General Framework for Estimating the Relative Pathogenicity of Human Genetic Variants", *Nat Genet.* 2014;46:310–5.

59. Sundaram, L *et al.*, "Predicting the Clinical Impact of Human Mutation with Deep Neural Networks", *Nat Genet.* 2018;50:1161–70.

60. Landrum, MJ *et al.*, "ClinVar: Improving Access to Variant Interpretations and Supporting Evidence", *Nucleic Acids Res.* 2018;46:D1062–7.

61. Riesselman, AJ, Ingraham, JB, Marks, DS, "Deep Generative Models of Genetic Variation Capture the Effects of Mutations", *Nat Methods.* 2018;15:816–22.

62. Chatterjee, S, Ahituv, N, "Gene Regulatory Elements, Major Drivers of Human Disease", *Annu Rev Genomics Hum Genet.* 2017;18:45–63.

63. Soemedi, R *et al.*, "Pathogenic Variants That Alter Protein Code Often Disrupt Splicing", *Nat Genet.* 2017;49:848–55.

64. Baeza-Centurion, P, Miñana, B, Schmiedel, JM, Valcárcel, J, Lehner, B, "Combinatorial Genetics Reveals A Scaling Law for the Effects of Mutations on Splicing", *Cell.* 2019; 176:549–63.

65. Kelley, DR, Reshef, YA, Bileschi, M, Belanger, D, McLean, CY, Snoek, J, "Sequential Regulatory Activity Prediction Across Chromosomes with Convolutional Neural Networks", *Genome Res.* 2018;28:739–50.

66. Alipanahi, B, Delong, A, Weirauch, MT, Frey, BJ., "Predicting the Sequence Specificities Of DNA- and-RNA-Binding Proteins by Deep Learning", *Nat Biotechnol.* 2015; 33:831–8.

67. Bernstein, BE *et al.*, "The NIH Roadmap Epigenomics Mapping Consortium", *Nat Biotechnol.* 2010;28:1045–8.

68. Zhou, J, Troyanskaya, OG, "Predicting Effects of Noncoding Variants with Deep Learning-Based Sequence Model", *Nat Methods.* 2015;12:931–4.

69. Zhou, J *et al.*, "Wholegenome Deep-Learning Analysis Identifies Contribution of Noncoding Mutations to Autism Risk", *Nat Genet.* 2019;51:973–80.

70. Zhou, J, Theesfeld, CL, Yao, K, Chen, KM, Wong, AK, Troyanskaya, OG, "Deep Learning Sequence-Based AB Initio Prediction of Variant Effects on Expression and Disease Risk", *Nat Genet.* 2018;50:1171–9.

71. Telenti, A *et al.*, "Deep Sequencing of 10,000 Human Genomes", *Proc Natl Acad Sci U S A.* 2016;113:11901–6.

72. Erikson, GA *et al.*, "Whole-Genome Sequencing of a Healthy Aging Cohort", *Cell.* 2016; 165:1002–11.

73. Köhler, S *et al.*, "Expansion of the Human Phenotype Ontology (HPO) Knowledge Base and Resources", *Nucleic Acids Res.* 2019;47:D1018–27.

74. Hsieh, T-C *et al.*, "PEDIA: Prioritization of Exome Data by Image Analysis", *Genet Med.* 2019. Available: https://doi. org/10.1038/s41436-019-0566-2.

75. Mobadersany *et al.*, "Predicting Cancer Outcomes from Histology and Genomics Using Convolutional Networks", *Proc Natl Acad Sci U S A*. 2018;115:E2970–9.
76. Bastarache, L. *et al.*, "Phenotype Risk Scores Identify Patients with Unrecognized Mendelian Disease Patterns", *Science*. 2018;359:1233–9.
77. Torkamani, A, Wineinger, NE, Topol, EJ., "The Personal and Clinical Utility of Polygenic Risk Scores", *Nat Rev Genet*. 2018;19:581–90.
78. Lello, L, Avery, SG, Tellier, L, Vazquez, AI. de los Campos, G, Hsu, SDH., "Accurate Genomic Prediction of Human Height", *Genetics*. 2018;210:477–97.
79. Lee, A *et al.*, "BOADICEA: A Comprehensive Breast Cancer Risk Prediction Model Incorporating Genetic and Nongenetic Risk Factors", *Genet Med*. 2019;21:1708–18.
80. Inouye, M *et al.*, "Genomic Risk Prediction of Coronary Artery Disease in 480,000 Adults", *J Am Coll Cardiol*. 2018;72:1883–93.
81. Topol, EJ, "High-Performance Medicine: The Convergence of Human and Artificial Intelligence", *Nat Med*. 2019; 25:44–56.
82. Lomas, N Google has used contract swaps to get bulk access terms to NHS patient data. TechCrunch. 2019; Available: https://techcrunch.com/2019/10/22/googlehas-used -contract-swaps-to-get-bulk-access-terms-to-nhs-patient-data/. Accessed 31 Oct 2019.
83. Vayena, E, Blasimme, A, Cohen, IG, "Machine Learning in Medicine: Addressing Ethical Challenges", *PLoS Med*. 2018;15:e1002689. Available: https://doi.org/10.1371/ journal.pmed.1002689.
84. Selvaraju, RR, Cogswell, M, Das, A, Vedantam, R, Parikh, D, Batra, D, "Grad-CAM: Visual Explanations from Deep Networks Via Gradient-Based Localization", In: International Conference on Computer Vision (ICCV): IEEE; 2017. pp. 618–26.
85. Olah, C, Mordvintsev, A, Schubert, L, "Feature Visualization: How Neural Networks Build up Their Understanding of Images", *Distill*. 2017;2:e7 https:// distill.pub/2017/ feature-visualization. Accessed 12 Aug 2019.
86. Mittelstadt, B, Russell, C, Wachter, S, "Explaining explanations in AI", In: FAT*, 2019. Proceedings of the 2019 Conference on Fairness, Accountability, and Transparency. Atlanta; 2019. p. 29, 279–31, 288. Available: https://doi.org/10.1145/ 3287560.3287574.
87. Doshi-Velez, F, Kim, B, "Towards a Rigorous Science of Interpretable Machine Learning", arXiv. 2017; arXiv:1702 08608.
88. Gianfrancesco, MA, Tamang, S, Yazdany, J, Schmajuk, G, "Potential Biases in Machine Learning Algorithms Using Electronic Health Record Data", *JAMA Intern Med*. 2018;178:1544–7.
89. Sirugo, G, Williams, SM, Tishkoff, SA, "The Missing Diversity in Human Genetic Studies", *Cell*. 2019;177:1080.
90. Lumaka, A *et al.*, "Facial Dysmorphism is Influenced by Ethnic Background of the Patient and of The Evaluator", *Clin Genet*. 2017;92:166–71.
91. Martin, AR, Kanai, M, Kamatani, Y, Okada, Y, Neale, BM, Daly, MJ, "Clinical use of Current Polygenic Risk Scores May Exacerbate Health Disparities", *Nat Genet*. 2019;51:584–91.
92. Bolukbasi, T, Chang, K-W, Zou, JY, Saligrama, V, Kalai, AT, "Man is to Computer Programmer as Woman is to Homemaker?", Debiasing word embeddings. In: Lee DD, Sugiyama M, Luxburg UV, Guyon I, Garnett R, editors. *Advances in neural information processing systems*, https://papers.nips.cc/paper/6228-man-is-tocomputer-programmer -as-woman-is-to-homemaker-debiasing-wordembeddings.pdf Accessed 31 Oct 2019.
93. Yarnell, CJ *et al.*, "Association between Immigrant Status and End-of-Life Care in Ontario", Canada. *JAMA*. 2017;318:1479–88.
94. Sohail, M *et al.*, "Polygenic Adaptation on Height s Overestimated Due to Uncorrected Stratification in Genome-Wide Association Studies", *Elife*. 2019;8. Available: https:// doi. org/10.7554/eLife.39702.

95. Chen, IY, Szolovits, P, Ghassemi, M, "Can AI help reduce disparities in general medical and mental health care?" *AMA J Ethics*. 2019;21:E167–79.

96. Sudlow, C *et al.*, "UK Biobank: An Open Access Resource for Identifying the Causes of a Wide Range of Complex Diseases of Middle and Old Age", *PLoS Med*. 2015;12:e1001779. https://doi.org/10.1371/journal.pmed.1001779.

97. Sankar, PL, Parker, LS., "The Precision Medicine Initiative's All of us Research Program: An Agenda for Research on its Ethical, Legal, and Social Issues", *Genet Med*. 2017;19:743–50.

4 AI and Genomes for Decisions Regarding the Expression of Genes

Abeedha Tu-Allah Khan and Haleema Qamar

CONTENTS

4.1 INTRODUCTION TO ARTIFICIAL INTELLIGENCE (AI)

The first milestone in the beginning and establishment of artificial intelligence (AI) was achieved by a British man known by the name of Alan Mathison Turing during the mid-20th century. He was born in London on June 23, 1912. He became an eminent and well-renowned logician and mathematician due to his contributions in mathematics, mathematical biology, philosophy of science, logistics, computer science, cryptanalysis, cognitive science, and artificial intelligence [1, 2].

The term artificial intelligence was coined from a combination of human intelligence and computer science to replicate biological intelligence into complex machines. The concept that devices will be trained to see, hear, think, and even reason like human beings has been termed "General AI." This concept has been pictured in many Hollywood movies, e.g., *Star Wars* in which a droid is programmed in such a way as to make it able to perform human-like protocols and etiquette. This level of training is not yet possible at present and thus represents somewhat unrealistic artificial intelligence goals. However, the idea of what is called "narrow AI" some

DOI: 10.1201/9781003126164-4

present-day technologies advancements in neurological sciences have been made. With the help of "narrow AI", human-like simple task performance has been done already, sometimes even better. Some of the most common examples of such tasks range from facial recognition, speech recognition, speech translation, text analysis, image analysis, and classification to applications such as language processing, vehicle automation, and in the field of medicine. Not only such technologies have been derived from AI techniques known commonly as machine and deep learning, but also these modern technologies are built on machine trainable features of human intelligence [3–5].

AI is playing a conspicuous and significant role in healthcare and medicine due to its various advantages. Some of these include advances in increasing computational powers and learning algorithms. Also, since the current data in the biological research field comes in large datasets – also known as big data – the analysis of this big data is considerably eased with the integration and use of AI. As a result, a 40% increase in the use of AI in the healthcare market has been observed. Apart from this, a rapid rise in a shift towards increasing use of high computing power systems is rooted in the use of graphical processing units (GPUs). The easy and widespread availability of GPUs enable the computer systems to become even faster than parallel processing performed by CPUs. Also, the availability and easy access to almost infinite cloud computing resources on-demand adds to their usefulness [6–9].

Continuous development is also present in the algorithms of AI themselves. For instance, the learning algorithms' precision and accuracy are increasing continuously with the arrival of new and better training datasets leading to better interactions with the data and ultimately to better insights into data analysis, diagnostic applications, patient treatment, and health outcomes. Along with this, the flooding of healthcare and research data is also pushing new interventions in AI applications with promising developments in the efficacies of the techniques [3, 10, 11].

Another somewhat more straightforward definition of AI is the simulation and replication of living beings' intelligence in non-living systems. In the context of clinical diagnostics, AI is defined as a computer system that is able to interpret health data in a correct way similar to that observed and done by humans. These applications for clinical data are often built upon AI frameworks to enable the fast and efficient interpretation of the large and complex clinical datasets. For the training of such AI systems, external health data are usually used with some properties. These properties include usual interpretation by human intelligence along with minimal processing before the data is exposed to the computer system. For example, while training an AI system with clinical image data, the information given to the machine includes interpretations by a human expert, which the computer system learns regarding those images. After the AI system has learned correctly, it can then perform interpretation tasks on new pictures of similar types and help in the identification of disease [12–18].

In this chapter, recent achievements in AI applications combined with their future improvements will be described with a particular focus on AI in genomics. First of all, clinical genomics will be discussed, covering relevant aspects of the field, including predictions of variant impacts, variant calling, annotation of variants, and mapping of phenotype to genotype. Next, the utility of AI in the context of modern genomics will be discussed, including conventional and next-generation sequencing techniques [12, 19, 20].

4.2 AI IN CLINICAL GENOMICS

The development of AI algorithms derives its inspiration from a desire to reproduce intelligence that exists in living organisms. The applications of AI in clinical genomic go one step further if their advantages are considered. This might be because human intelligence becomes error-prone when combined with statistical methodologies and leads to some impracticability in performing such tasks. However, AI is a liberator when analyzing clinical genomic data. Further advantages include its applicability for predictions of phenotypes. Now, a detailed description of these individual sub-types of clinical genomics data analysis via AI will be given [21, 22].

4.2.1 VARIANT CALLING

One of the most perplexing and essential genomics missions is the accurate identification of the variants in DNA sequence. The process of finding differences in nucleotides between a given genome as compared to another genome, often known as the reference genome, is known as variant calling (Figure 4.1). The accuracy and efficiency of calling variants are indispensable since the downstream applications are the correct detection of the disease and pinpointing the phenotypic variances arising as a result of genotype differences. Another reason why variant calling demands extreme accuracy is that the clinical interpretation of the genetic variants is highly sensitive to identifying the genetic variants individually and that too among millions of variants present in a genome [23–25].

There are a variety of variant-calling tools available. However, they are predisposed to systematic errors which are concomitant with differences in preparation

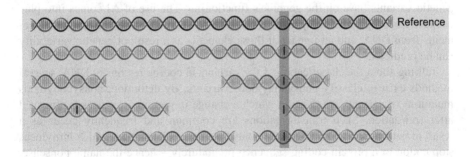

FIGURE 4.1 Variant calling/detection: The variant calling or detection process involves the alignment of the target sequence reads (all DNA molecules below the reference strand) with a reference sequence (top DNA strand). After this, nucleotide variants, insertions, or deletions are detected, and these variants are then annotated in comparison with the reference. For simplicity, only single nucleotide polymorphisms are represented here in red.

methods of the samples; technology or machine sensitivities used for sequencing genomes; subtleties in the context of sequences; and somatic mosaicism, which sometimes poses unpredictable biological influences. Some of the statistical methods have been developed to address these issues, of which most common are population-level dependencies and strand-bias techniques. Although the utilization of these methods resulted in highly accurate results, at the same time, the biased errors were also increased [26–29].

Some studies have shown that AI algorithms trained using a single genome along with a standard gold reference are more efficient in variant calling and outperform other algorithms. For example, DeepVariant is an algorithm that has been trained on a convolutional neural network (CNN)-based variant caller without the added complications of sequencing platforms or genomics knowledge. It has been shown to outpace other similar variant caller programs. The primary reason for its improvised efficiency was attributed to CNN's ability in the identification and interpretation of complexities present in the sequencing data. Recent research is also suggestive of deep learning capabilities in revolutionizing base calling, which ultimately leads to variant calling and identification. It is especially true for nanopore sequencing technologies to overcome their error-prone nature.

4.2.2　VARIANT CLASSIFICATION AND ANNOTATION OF GENOMES

The process of genome sequencing is explained using two methods known as variant classification and genome annotation. In the former, the genes are classified according to the mutations or pathogenic sequences present, while the latter involves naming the genes present.

4.2.2.1　Coding Mutants/Variants

Once the genomic data has been passed through variant calling algorithms, the next principal aim is to identify and interpret gene variants' occurrence. This is done via combining knowledge from the prior information on the impact that particular genetic variants have on the genome's functioning. The use of AI comes into play because these algorithms play an essential role in the prediction of the functional elements from DNA, and an impact of those elements as a result of genetic variability can be predicted.

Talking about the classification of the variants in coding regions of DNA, several methods exist to classify nonsynonymous variants. By definition, nonsynonymous mutations or variants are those in which a change in protein sequences is observed after translation. Such transformations are common and frequently occur as a result of natural selection. The same holds for nonsense mutation, which introduces stop codon in a protein-coding sequence prematurely – hence the name nonsense. Contrastingly, synonymous mutations do not affect the functionality of the genes, and silence is not transferred into the protein, i.e., the sequence of the protein does not change. Hence such mutations are also evolutionarily neutral [30–32].

While some methods have been developed to classify nonsynonymous mutations per se, others are integrated into meta predictors built on deep learning techniques.

These meta predictors are prediction models trained to process the predictions and merge other predictions made by several other predictors. This assimilation results in the outperformance in predictions at levels of individual forecasts and the integration of predictions with other machine learning methods such as regression. For a better explanation of this, let us consider CADD's example – the combined annotation-dependent depletion approach. Typically used to classify coding variants, this algorithm is built on a combination of predictive features for deleterious genetic mutations. However, there is an extension of this algorithm – DANN – which is based on the program's deep learning training. As it combines the primary input feature of CADD with deep neural network training, DANN results in the demonstration of improved performance, suggesting deep learning to be a superior approach for the integration of features of deleterious mutation predictions [12, 33–35].

Other AI methods for direct predictions from sequences of DNA or proteins exist with fewer handcrafted features. One such algorithm is Primate AI. It is trained on CNNs. Thus, the advantage of its outperformance lies in the augmentation of data with cross-species information about pathogenic variants of disease as compared to the training, which relies only on sequence alignment methods [36]. Besides this, a new algorithm has been presented for the prediction of deleterious missense mutations. It is built on a new machine learning supervised model known by the name of MISTIC (MISsense deleTeriousness predICtor). It is based on two algorithms whose functioning are complementary to each other, i.e., logistic regression and the random forest approach. The advantage of this AI algorithm is its ability to distinguish benign missense mutations from harmful ones. This program training is based on the selection of missense features of coding sequence along with changes in the bio- and physiochemical properties of the amino acids as a result. When the performance of the MISTIC with other prediction tools – PrimatAI, ClinPred, M-CAP, Eigen, FATHMM-XF, and REVEL – was done, the best results were obtained for the prediction and ranking of deleterious variants of the coding DNA sequences [37–40].

4.2.2.2 Non-Coding Mutants/Variants

One of the most significant challenges open in human genomics is the precise identification and prediction of pathogenic variations present in the non-coding regions of the DNA by computational means. A considerable candidate in the functional interpretation of non-coding variants in DNA is enhancers. Enhancers are short DNA sequences that function to promote the transcription of a gene when specific proteins bind to them and thus function together to enhance the process [41]. An efficient example of the non-coding regions in the pathogenicity of a disease can be seen by considering a gene known as nitric oxide synthase one adapter protein (NOS1AP) alias carboxyl-terminal PDZ ligand of neuronal nitric oxide synthase protein (CAPON). This gene is present in chromosome 1q in the human genome. It is well known for its association with cardiac repolarization as well as QT interval (QT interval is the time taken from the start of Q wave till the end of T wave typically representing ventricular de- and repolarization) [42–45].

Despite the long-known association of NOS1AP with cardiac function, the underlying mechanism remained unclear until recently. A recent study has identified

210 common variants posing a risk to the QT interval variation. In this study, the researchers utilized resequencing techniques combined with fine-mapping via a genome-wide association study (GWAS) of the gene's locus. All of these risk variants turned out to be present in the non-coding region of the DNA. Also, when the analysis of the suppressor/enhancer regions of the respective gene was done, it was found that 12 variants of the cardiac phenotype were associated with the DNase I hypersensitivity, which further aided in the identification of another upstream variant in the enhancer region [46, 47]. Similarly, splicing defects in the genes constitute about 10% of genetic variations in rare pathogenic diseases [48].

Because of difficulties in identifying the splicing defects due to the complex nature of the splicing and DNA interactions between intronic and exonic regions, an algorithm has been designed known as Splice AI. It is a 32-layer deep neural network capable of predicting both splicing types, i.e., either non-canonical or canonical, from just the sequence data of intron-exon junction. It is also remarkable in boosting the prediction accuracy by 57% despite integrating a short window size of 80 nucleotides. This accuracy became possible via the utilization of long-range sequence information [48–51].

4.3   AI IN GENE EXPRESSION DATA ANALYSIS

The data constituting the relative or absolute abundance of mRNA or transcripts of a biological sample are generally known as gene expression data. Although the gene expression studies' primary aim is to objectively measure the number of mRNA molecules of a particular gene. These mRNAs are sometimes referred to as objective units; however, it is rarely possible to measure them in literal sense. Instead, gene expression measurements are qualitative most of the time, and even when quantitative measurements are done, they are associated with often unknown and high amounts of noise. This particular factor must be taken into account while analyzing expression data or designing artificial intelligence programs to carry out the same [52, 53].

Of the many technologies utilized to measure and study gene expression, some of the most common during the past few decades have been northern blots, quantitative real-time PCR (RT-PCR), serial analysis of gene expression (SAGE), and microarrays. Of these, the best method has been the microarrays for an extensive library of transcripts. Besides microarrays being consistently used in the modern era, RNA sequencing is also becoming a popular choice of model for gene expression analysis studies due to its associated benefits being more significant than microarrays. However, the analysis of both of these high-throughput methods has been built upon technologies based on the core principles of artificial intelligence methods [54–58].

4.3.1   DIMENSIONALITY REDUCTION

Despite the same number of samples in the integrative analysis of big biological data, the number of study features or variables is generally increased. It can be attributed to the reason that the multiple platform measurements often belong to the same sample. It is also known as the dimensionality curse in terms of artificial intelligence language. Whenever p >> n, this raises the need for programs efficient enough to tackle the problem of dimensionality. One of the most apparent consequences of increased dimensionality is an increased vulnerability to overfitting problems in most machine learning methods. It can also result in high accuracy concerning the training data but has low efficiency in generalizing the unseen test data. Another reason for this is the coverage of even smaller input feature space fractions by the same samples. Although new features carry new information too, the dimensionality curse results in the out-weightage of the latest information benefit coming along with new features [59–62]. Dimensionality reduction is typical in transcriptomics analysis. The most common techniques for this are feature extraction and feature selection. The essential purpose of feature extraction is the projection of the data to lower its dimensional space. In contrast, feature selection's primary function is the identification of a relevant subset of data from the original features for dimensionality reduction (Figure 4.2) [61, 63, 64].

4.3.1.1 Feature Extraction

The main advantages of feature extraction of gene expression data include the facilitation of data visualization and exploration, profiling of hidden factors, and compression analysis. The most common feature extraction method in use is principal component analysis (PCA) from dimensionality reduction of high dimensional features of the data. It functions to transform the data orthogonally and produces linear uncorrelated principal components. The maximal variance of the datasets is also captured in the principal components. The decomposition matrices resulting from the principal component analysis have values present in both the positive and negative ranges. Also, when PCA is combined with clustering, an additional benefit of exploratory data analysis is achieved. It becomes especially useful for sub-groups'

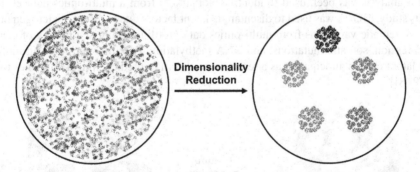

FIGURE 4.2 Dimensionality reduction: A process in gene expression analysis includes a reduction in a large cohort of variables – in this case, genes – into a smaller number of variables/groups based on similarity. The circle on the left side indicates the data distributed in all directions, which is difficult to process and leads to the poor performance of the machine learning algorithms. The right circle represents the data after dimensionality reduction has been made, and variables (or genes) have been arranged into clusters based on similarities or other common features.

visualization in a dataset that is otherwise uninterpretable under high dimensionality conditions (Figure 4.3) [65–69].

Figure 4.3 divided into six features by the feature extraction algorithms. These extracted features will then be used to train the machine for further learning.

Another method of feature extraction is known as non-negative matrix factorization (NMF). It produces two non-negative matrices as a result of dimensionality reduction. The size of the product matrices also matches to the original matrix and product matrices are also non-negative. Since the resulting matrices from non-negative matrix factorization have positive values only, the original data's representation is done by combining the latent variables additively. One example of NMF feature extraction algorithm is the t-distributed stochastic neighbor embedding (t-SNE) algorithm. It is built on a non-linear method for dimensionality reduction and performs better when non-linear relationships are present in the data. Also, the construction of joint probability distributions from the individual data points is based on their similarities. There is a minimal divergence between the joint probabilities after dimensionality reduction and the original high-dimensional data [70–74].

Feature extraction methods are often utilized in studies where group labels of data cannot carry out an unsupervised and integrative analysis. Feature extraction using machine learning-based approaches is particularly useful as they facilitate the discovery of sub-groups specific to a particular disease in studies involving multi-omics methodologies. Many feature extraction methods based on PCA have been put forward for the integrative exploratory analysis of omics datasets [75–78].

Let us now see some examples of the unsupervised, integrative analysis techniques, and how they have been utilized and proved to be useful for the analysis of multi-omics data. One of them has been proposed relatively recently, known by the name of multi-omics factor analysis (MOFA). A generalization of principal component analysis has been used to identify biomarkers from a multi-omics dataset. In this study, MOFA was used to disentangle latent factors due to disease heterogeneity or systematic variations from multi-omics data profiling comprised mainly of gene expression, somatic mutations, and DNA methylation. Also, the periodic variations or latent factors anticipated, as a result, were shown to predict clinical outcomes too [79–81].

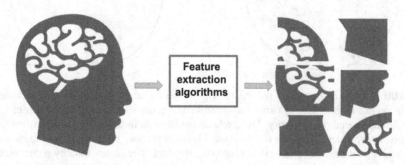

FIGURE 4.3   Feature extraction is another method of dimensionality reduction.

Another extension of the principal component analysis, known as joint and individual variation explained (JIVE), was proposed to identify variations at both individual and combined levels between gene expression and miRNA data. It is also based on integrating exploratory data analysis approaches and works via the decomposition of a dataset into three main terms. These are two approximation terms with low ranks with each designed to capture joint and individual structures of the data separately, and the third term for capturing residual noise [82–85]. Similarly, for the integration of gene and protein expression datasets, multiple co-inertia analysis (MCIA) was proposed. It is mainly based on PCA analysis of each dataset separately, ultimately reducing their dimensional spaces to similar levels for exploratory data analysis. One advantage of this lies in the better interpretation of the biological pathways via efficiently combining features of both genes and proteins, which have been transformed earlier to a similar scale [86, 87].

4.3.1.2 Feature Selection

About the gene expression datasets, the most common objective is to classify the samples correctly. In this context, the process of feature selection can be explained as a route to find those genes with a minimum probability of inducing a classification error in the samples. In other words, their selection should ideally result in the maximization of the samples' classification accuracy (Figure 4.4). Most of the feature selection algorithms have been designed based on the same principle. Some of the real-world applications harboring the feature selection techniques include the analysis of large-scale biological data; information retrieval and text classification; spectroscopy and its data analysis; quantitative structure-activity relationship (QSAR) modeling in drug designing; and miRNA, mRNA, proteomics, whole-exome sequencing, and DNA-methylation analysis in biological omics data analysis [88–98].

Various classification strategies have been devised and approached for choosing an appropriate method for unpaired feature selection for a particular scenario.

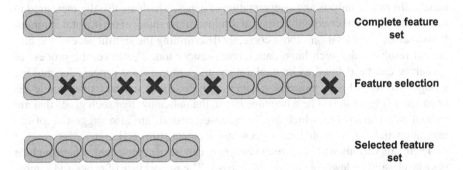

FIGURE 4.4 In feature selection, the features of interest are selected manually or automatically, which are most relevant to the data which has to be kept. In this figure, the boxes with circles in them represent interest, and green boxes are unnecessary features. Therefore, in the final set of features, unnecessary ones have been omitted while keeping the essential ones.

The classification methods used in some standard techniques include embedded, wrapper, and filter methods mainly named based on the classifiers integrated into their programs. The filter approach is frequently applied to high-dimensional data analysis, such as microarray or RNA-seq data. It functions to assess the relevant features of the data depending on the intrinsic properties of designed filters. This approach is developed on the separation of classifier construction from feature selection. The wrapper approach is based on the gradual optimization of or searching for the selected features with an evaluation of the classification performance along the process to satisfy a specific set criterion. The embedded approach is computationally less intensive than the wrapper approach and is based on the classifier classification and feature selection embedding. It bears an additional advantage of interacting with models designed for classification. However, the best results have been observed in ensemble feature selection approaches in which the algorithm is built upon the integration of multiple models into one. The most useful advantage of this approach lies in its ability to handle stability issues which have been lacking in the previous methods [99–105].

4.3.2 CLUSTERING

While working with gene expression data, a critical computational and statistical manipulation is the clustering of genes and samples. It is a standard technique used to identify clusters of genes that are closely related to each other. Since its advent, DNA microarray technology has offered great opportunities for the simultaneous measurement of all the genes at a given time in a particular sample. However, in the current era, the technology of RNA sequencing has offered additional benefits as compared to microarrays in that next-generation RNA sequencing provides ways to measure the expression of splice variants and novel transcripts as well, besides fundamental differential gene expression analysis. As all of these high-throughput techniques typically result in large amounts of data, this also poses a significant challenge for data mining fields. Some of the most common of these challenges include the rapid analysis and interpretation of genes, which are usually measured in thousands with further complications arising from the many experimental samples. It also raises considerations about correctly determining the significance of the biological results under such huge data circumstances too. Therefore, the process of clustering can be defined as an exploratory and unsupervised process for making groups / clusters of gene expression data mainly based on similar expression levels or patterns (Figure 4.5). The advantage lies in the indication that such genes that are present in a cluster, i.e., which are being co-expressed, are also suggestive of co-regulation and mutual biological process involvement [106–111].

There are various ways in which clustering can be accomplished. These include gene level, sample, level, or time variable basis. The importance of genes and sample clusters lies in the exhibition of similar expressions across different conditions in gene-based clustering. In contrast, similarity across the samples concerning relevant gene expression is observed in sample clustering approaches. For each clustering type, the type is referred to as an object, e.g., the items in gene- and sample-based

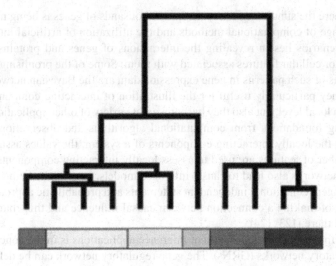

FIGURE 4.5 Gene clustering, a process of grouping similar genes into clusters. For example, two similar genes have been clustered into a box on right, while the second block from left is indicative of a cluster made up of three genes. The horizontal lines represent the distance and hence the divergence in the similarities of the sets.

clustering techniques are genes and samples, respectively. However, clustering's peculiarity and specificity are different for different gene expression datasets depending on their characteristics [110, 112–114].

Two types of clustering can be performed on gene expression data, i.e., partial clustering and complete clustering. The former of these, partial clustering, bears more suitability for gene expression studies as these are often comprised of some irrelevant genes or samples too. Most of the time in gene expression data, some genes represent noise, resulting in a correspondingly lesser impact on the outcome; hence partial clustering prevents the allotment of such noisy genes to particular clusters, thus also preventing their irrelevant contribution. In this way, it also helps preserve a sub-group in which one is particularly interested in avoiding the involvement of unrelated genes [110, 115–118]. Some other categorizations of clustering include problematic or overlapping types. The former type includes clustering to a single cluster during the clustering operation and the output. While in the latter class, the collections are assigned for the membership degrees in several clusters about individual input genes. However, the overlapping type gets transformed into the hard type when the individual genes get assigned with the membership having a dominant degree(s) [119–122].

4.3.3 Bayesian Networks

The Bayesian network represents a graphical model built on probabilistic approaches. Known by different names such as Bayes, decision, and belief network, it is representative of variables' set and conditional dependencies. In the context of gene expression studies where the simultaneous expression of thousands of genes is being measured, the challenge of computational methods and the utilization of artificial intelligence in such scenarios lies in revealing the interactions of genes and proteins and the biological or cellular features associated with them. Some of the promising tools for the analysis of such patterns in gene expression data are the Bayesian networks. Not only are they particularly useful for the illustration of interacting components of a system at a local level, but also they harbor a great variety of other applications based on learning foundations from computational algorithms and observations. While describing the locally interacting components of a system, the values assigned to a small number of features are used to assess locally interacting components' values. Bayesian networks also lead to causal inference models, i.e., despite being designed under stringent conditional independent statements and probabilistic approaches, the possibility of making a connection between causal influence and this characterization is still there [123, 124].

One of the most common Bayesian inference applications is in the conception of gene regulatory networks (GRNs). The gene regulatory network can be defined as a hierarchical network of genes, their respective regulatory proteins, and their interactions. These regulatory networks play essential roles in the mediation of signaling pathways and the cellular functions associated with them. The special significance of the gene-related networks lies in their accurate inference concerning specific diseases, thereby identifying the potential targets of that disease and targeted therapeutic interventions. With the growing advancements in DNA analysis technology, an investigation of the non-coding DNA is increasingly associated with its robust effects on the transcription of genes [125–128].

Several measurements from multiple data types are inevitable for the proper analysis of gene regulatory networks. Some of the most trivial of these data types are gene expression, transcription factors, chromatin binding, methylation, and histone modifications. The main challenge of the inference of GRNs lies in the inherent complexity of the biological systems, i.e., the existence of hundreds and thousands of proteins and genes and the variance of their interactive relationships to not only the types of tissues and diseases but also for the individual cells as well. Hence, the usefulness of Bayesian inference for this type of large-scale data integration is due to its flexibility. The generation of multiple datasets under similar conditions results in more accessible and robust final estimation across numerous datasets when distribution learning methods are utilized [129–131].

4.4 CONCLUSION

AI systems are increasingly surpassing state-of-the-art methods for various clinical diagnostics applications and have already gained clearance from the FDA. The main driving factor for this productivity surge has become possible with the progressive availability of large datasets to train AI models, advances in AI algorithms, and the speed acceleration achieved with graphical processing units in training systems. In the current times, extracting phenotypic data from image processing and downstream genetic analysis is the most promising application in clinical genomics. Other applications of deep-learning algorithms of AI include genome annotation, variant calling, and genotype to phenotype predictions. Modern next-generation sequencing technology allows for large-scale data uniformity and generation. However, the data collection appears to be slow, variable, and highly expensive. The machine learning models help identify the gene biomarkers for various diseases, of which cancer detection is among the most common ones. It represents a significant step in individual diagnostics and shall aid in the development and implementation of personalized therapies.

BIBLIOGRAPHY

1. Encyclopædia Britannica. *Alan Turing*. June 19, 2019 October 05, 2020]; Available from: https://www.britannica.com/biography/Alan-Turing.
2. Encyclopædia Britannica. *Alan Turing and the beginning of AI*. August 11, 2020 October 05, 2020]; Available from: https://www.britannica.com/technology/artificial-intelligence.
3. Ahuja, A.S., *The impact of artificial intelligence in medicine on the future role of the physician*. PeerJ, 2019. **7**: p. e7702-e7702.
4. Davenport, T. and R. Kalakota, *The potential for artificial intelligence in healthcare*. Future Healthcare Journal, 2019. **6**(2): p. 94–98.
5. Nvidia. *What's the difference between artificial intelligence, machine learning and deep learning?* July 29, 2016; Available from: https://blogs.nvidia.com/blog/2016/07/29/whats-difference-artificial-intelligence-machine-learning-deep-learning-ai/.
6. Lopes, N. and B. Ribeiro, *An evaluation of multiple feed-forward networks on GPUs*. Int J Neural Syst, 2011. **21**(1): p. 31–47.
7. Navale, V. and P.E. Bourne, *Cloud computing applications for biomedical science: A perspective*. PLoS Computational Biology, 2018. **14**(6): p. e1006144-e1006144.
8. Vinuesa, R., et al., *The role of artificial intelligence in achieving the Sustainable Development Goals*. Nature Communications, 2020. **11**(1): p. 233–233.
9. Alexander, A., et al., *An Intelligent Future for Medical Imaging: A Market Outlook on Artificial Intelligence for Medical Imaging*. J Am Coll Radiol, 2020. **17**(1 Pt B): p. 165–170.
10. Bali, J., R. Garg, and R.T. Bali, *Artificial intelligence (AI) in healthcare and biomedical research: Why a strong computational/AI bioethics framework is required?* Indian Journal of Ophthalmology, 2019. **67**(1): p. 3–6.
11. Craft, JA, 3rd, *Artificial Intelligence and the Softer Side of Medicine*. Missouri Medicine, 2018. **115**(5): p. 406–409.
12. Dias, R. and A. Torkamani, *Artificial intelligence in clinical and genomic diagnostics*. Genome Medicine, 2019. **11**(1): p. 70–70.
13. Hosny, A., et al., *Artificial intelligence in radiology*. Nature reviews. Cancer, 2018. **18**(8): p. 500–510.
14. Pesapane, F., M. Codari, and F. Sardanelli, *Artificial intelligence in medical imaging: threat or opportunity? Radiologists again at the forefront of innovation in medicine*. European Radiology Experimental, 2018. **2**(1): p. 35–35.
15. Gyftopoulos, S., et al., *Artificial Intelligence in Musculoskeletal Imaging: Current Status and Future Directions*. AJR. American Journal of Roentgenology, 2019. **213**(3): p. 506–513.
16. England, J.R. and P.M. Cheng, *Artificial Intelligence for Medical Image Analysis: A Guide for Authors and Reviewers*. AJR Am J Roentgenol, 2019. **212**(3): p. 513–519.

17. Robertson, S., et al., *Digital image analysis in breast pathology-from image processing techniques to artificial intelligence.* Transl Res, 2018. **194**: p. 19–35.

18. Özdemir, V., *The Big Picture on the "AI Turn" for Digital Health: The Internet of Things and Cyber-Physical Systems.* Omics, 2019. **23**(6): p. 308–311.

19. Nagarajan, N., et al., *Application of Computational Biology and Artificial Intelligence Technologies in Cancer Precision Drug Discovery.* BioMed Research International, 2019. **2019**: p. 8427042–8427042.

20. Álvarez-Machancoses, Ó., et al., *On the Role of Artificial Intelligence in Genomics to Enhance Precision Medicine.* Pharmacogenomics and Personalized Medicine, 2020. **13**: p. 105–119.

21. Jiang, F., et al., *Artificial intelligence in healthcare: past, present and future.* Stroke and Vascular Neurology, 2017. **2**(4): p. 230.

22. Phan, N.N., A. Chattopadhyay, and E.Y. Chuang, *Role of artificial intelligence in integrated analysis of multi-omics and imaging data in cancer research.* 2019, 2019. **8**(8): p. E7–E10.

23. Bohannan, Z.S. and A. Mitrofanova, *Calling Variants in the Clinic: Informed Variant Calling Decisions Based on Biological, Clinical, and Laboratory Variables.* Computational and Structural Biotechnology Journal, 2019. **17**: p. 561–569.

24. Chen, J., et al., *Systematic comparison of germline variant calling pipelines cross multiple next-generation sequencers.* Sci Rep, 2019. **9**(1): p. 9345.

25. Luo, R., et al., *A multi-task convolutional deep neural network for variant calling in single molecule sequencing.* Nature Communications, 2019. **10**(1): p. 998.

26. Li, H., *Toward better understanding of artifacts in variant calling from high-coverage samples.* Bioinformatics, 2014. **30**(20): p. 2843–51.

27. Garrison, E. and G.J.A.P.A. Marth, *Haplotype-based variant detection from short-read sequencing.* 2012.

28. DePristo, M.A., et al., *A framework for variation discovery and genotyping using next-generation DNA sequencing data.* Nat Genet, 2011. **43**(5): p. 491–8.

29. Hwang, S., et al., *Systematic comparison of variant calling pipelines using gold standard personal exome variants.* Sci Rep, 2015. **5**: p. 17875.

30. Chu, D. and L. Wei, *Nonsynonymous, synonymous and nonsense mutations in human cancer-related genes undergo stronger purifying selections than expectation.* BMC Cancer, 2019. **19**(1): p. 359–359.

31. Nielsen, R. and D.M. Weinreich, *The age of nonsynonymous and synonymous mutations in animal mtDNA and implications for the mildly deleterious theory.* Genetics, 1999. **153**(1): p. 497–506.

32. Loewe, L., et al., *Estimating selection on nonsynonymous mutations.* Genetics, 2006. **172**(2): p. 1079–1092.

33. Quang, D., Y. Chen, and X. Xie, *DANN: a deep learning approach for annotating the pathogenicity of genetic variants.* Bioinformatics, 2015. **31**(5): p. 761–3.

34. Kircher, M., et al., *A general framework for estimating the relative pathogenicity of human genetic variants.* Nat Genet, 2014. **46**(3): p. 310–5.

35. Rentzsch, P., et al., *CADD: predicting the deleteriousness of variants throughout the human genome.* Nucleic Acids Res, 2019. **47**(D1): p. D886–d894.

36. Sundaram, L., et al., *Predicting the clinical impact of human mutation with deep neural networks.* Nat Genet, 2018. **50**(8): p. 1161–1170.

37. Chennen, K., et al., *MISTIC: A prediction tool to reveal disease-relevant deleterious missense variants.* PloS one, 2020. **15**(7): p. e0236962-e0236962.

38. Goldman, S.A. and MKJML Warmuth, *Learning binary relations using weighted majority voting.* 1995. **20**(3): p. 245–271.

39. Breiman, L.J.M.l., *Random forests.* 2001. **45**(1): p. 5–32.

40. Collins, M., RE Schapire, and YJML Singer, *Logistic regression, AdaBoost and Bregman distances.* 2002. **48**(1–3): p. 253–285.

41. Pennacchio, L.A., et al., *Enhancers: five essential questions.* Nature Reviews. Genetics, 2013. **14**(4): p. 288–295.

42. Postema, P.G. and A.A.M. Wilde, *The measurement of the QT interval.* Current Cardiology Reviews, 2014. **10**(3): p. 287–294.

43. Kapoor, A., et al., *An enhancer polymorphism at the cardiomyocyte intercalated disc protein NOS1AP locus is a major regulator of the QT interval.* American Journal of Human Genetics, 2014. **94**(6): p. 854–869.

44. Chang, K.-C., et al., *Nitric Oxide Synthase 1 Adaptor Protein, an Emerging New Genetic Marker for QT Prolongation and Sudden Cardiac Death.* Acta Cardiologica Sinica, 2013. **29**(3): p. 217–225.

45. Semsarian, C., J. Ingles, and A.A. Wilde, *Sudden cardiac death in the young: the molecular autopsy and a practical approach to surviving relatives.* Eur Heart J, 2015. **36**(21): p. 1290–6.

46. Arking, DE, et al., *A common genetic variant in the NOS1 regulator NOS1AP modulates cardiac repolarization.* Nat Genet, 2006. **38**(6): p. 644–51.

47. Kapoor, A., et al., *An enhancer polymorphism at the cardiomyocyte intercalated disc protein NOS1AP locus is a major regulator of the QT interval.* Am J Hum Genet, 2014. **94**(6): p. 854–69.

48. Soemedi, R., et al., *Pathogenic variants that alter protein code often disrupt splicing.* Nat Genet, 2017. **49**(6): p. 848–855.

49. Baeza-Centurion, P., et al., *Combinatorial Genetics Reveals a Scaling Law for the Effects of Mutations on Splicing.* Cell, 2019. **176**(3): p. 549-563.e23.

50. Chatterjee, S. and N. Ahituv, *Gene Regulatory Elements, Major Drivers of Human Disease.* Annu Rev Genomics Hum Genet, 2017. **18**: p. 45–63.

51. Jaganathan, K., et al., *Predicting Splicing from Primary Sequence with Deep Learning.* Cell, 2019. **176**(3): p. 535-548.e24.

52. Singh, K.P., et al., *Mechanisms and Measurement of Changes in Gene Expression.* Biological Research for Nursing, 2018. **20**(4): p. 369–382.

53. Gibcus, J.H. and J. Dekker, *The context of gene expression regulation.* F1000 Biology Reports, 2012. **4**: p. 8–8.

54. Tarasov, KV, et al., *Serial Analysis of Gene Expression (SAGE): a useful tool to analyze the cardiac transcriptome.* Methods Mol Biol, 2007. **366**: p. 41–59.

55. Tarca, A.L., R. Romero, and S. Draghici, *Analysis of microarray experiments of gene expression profiling.* American Journal of Obstetrics and Gynecology, 2006. **195**(2): p. 373–388.

56. He, SL and R. Green, *Northern blotting.* Methods in Enzymology, 2013. **530**: p. 75–87.

57. Bachman, J., *Reverse-transcription PCR (RT-PCR).* Methods Enzymol, 2013. **530**: p. 67–74.

58. Kukurba, K.R. and S.B. Montgomery, *RNA Sequencing and Analysis.* Cold Spring Harbor protocols, 2015. **2015**(11): p. 951–969.

59. De Meulder, B., et al., *A computational framework for complex disease stratification from multiple large-scale datasets.* BMC Syst Biol, 2018. **12**(1): p. 60.

60. Cavill, R., et al., *Transcriptomic and metabolomic data integration.* Brief Bioinform, 2016. **17**(5): p. 891–901.

61. Wang, L., Y. Wang, and Q. Chang, *Feature selection methods for big data bioinformatics: A survey from the search perspective.* Methods, 2016. **111**: p. 21–31.

62. Min, S., B. Lee, and S. Yoon, *Deep learning in bioinformatics.* Brief Bioinform, 2017. **18**(5): p. 851–869.

63. Hira, ZM and D.F. Gillies, *A Review of Feature Selection and Feature Extraction Methods Applied on Microarray Data.* Adv Bioinformatics, 2015. **2015**: p. 198363.

64. Guyon, I. and A.J.J.o.m.l.r. Elisseeff, *An Introduction to Variable and Feature Selection.* 2003. **3**(Mar): p. 1157–1182.

65. Tsuyuzaki, K., et al., *Benchmarking principal component analysis for large-scale single-cell RNA-sequencing.* Genome Biology, 2020. **21**(1): p. 9.

66. Marini, F. and H. Binder, *pcaExplorer: an R/Bioconductor package for interacting with RNA-seq principal components.* BMC Bioinformatics, 2019. **20**(1): p. 331.

67. Son, K., et al., *A Simple Guideline to Assess the Characteristics of RNA-Seq Data.* BioMed Research International, 2018. **2018**: p. 2906292–2906292.

68. Black, M.H. and R.M. Watanabe, *A principal components-based clustering method to identify variants associated with complex traits.* Human Heredity, 2011. **71**(1): p. 50–58.

69. Wang, K., W. Wang, and M. Li, *A brief procedure for big data analysis of gene expression.* Animal Models and Experimental Medicine, 2018. **1**(3): p. 189–193.

70. Maaten, L.v.d. and G.J.J.o.m.l.r. Hinton, *Visualizing data using t-SNE.* 2008. **9**(Nov): p. 2579–2605.

71. Hinton, G.E. and R.R. Salakhutdinov, *Reducing the dimensionality of data with neural networks.* Science, 2006. **313**(5786): p. 504–7.

72. Wang, Y., H. Yao, and S.J.N. Zhao, *Auto-encoder based dimensionality reduction.* 2016. **184**: p. 232–242.

73. Cieslak, M.C., et al., *t-Distributed Stochastic Neighbor Embedding (t-SNE): A tool for eco-physiological transcriptomic analysis.* Mar Genomics, 2020. **51**: p. 100723.

74. Linderman, GC, et al., *Fast interpolation-based t-SNE for improved visualization of single-cell RNA-seq data.* Nat Methods, 2019. **16**(3): p. 243–245.

75. Meng, C., et al., *Dimension reduction techniques for the integrative analysis of multi-omics data.* Brief Bioinform, 2016. **17**(4): p. 628–41.

76. Jain, Y., S. Ding, and J. Qiu, *Sliced inverse regression for integrative multi-omics data analysis.* Stat Appl Genet Mol Biol, 2019. **18**(1).

77. Altmäe, S., et al., *Guidelines for the design, analysis and interpretation of 'omics' data: focus on human endometrium.* Hum Reprod Update, 2014. **20**(1): p. 12–28.

78. Meng, C. and A. Culhane, *Integrative Exploratory Analysis of Two or More Genomic Datasets.* Methods Mol Biol, 2016. **1418**: p. 19–38.

79. Argelaguet, R., et al., *Multi-Omics Factor Analysis-a framework for unsupervised integration of multi-omics data sets.* Mol Syst Biol, 2018. **14**(6): p. e8124.

80. Argelaguet, R., et al., *MOFA+: a statistical framework for comprehensive integration of multi-modal single-cell data.* Genome Biol, 2020. **21**(1): p. 111.

81. Hasin, Y., M. Seldin, and A. Lusis, *Multi-omics approaches to disease.* Genome Biol, 2017. **18**(1): p. 83.

82. Lock, E.F., et al., *JOINT AND INDIVIDUAL VARIATION EXPLAINED (JIVE) FOR INTEGRATED ANALYSIS OF MULTIPLE DATA TYPES.* Ann Appl Stat, 2013. **7**(1): p. 523–542.

83. O'Connell, M.J. and E.F. Lock, *R.JIVE for exploration of multi-source molecular data.* Bioinformatics, 2016. **32**(18): p. 2877–9.

84. Kuligowski, J., et al., *analysis of multi-source metabolomic data using joint and individual variation explained (JIVE).* Analyst, 2015. **140**(13): p. 4521–9.

85. Yu, Q., et al., *JIVE integration of imaging and behavioral data.* Neuroimage, 2017. **152**: p. 38–49.

86. Meng, C., et al., *A multivariate approach to the integration of multi-omics datasets.* BMC Bioinformatics, 2014. **15**: p. 162.

87. Min, EJ and Q. Long, *Sparse multiple co-Inertia analysis with application to integrative analysis of multi -Omics data.* BMC Bioinformatics, 2020. **21**(1): p. 141.

88. Bermingham, M.L., et al., *Application of high-dimensional feature selection: evaluation for genomic prediction in man.* 2015. **5**: p. 10312.

89. Liu, H.-J., et al., *Predicting novel salivary biomarkers for the detection of pancreatic cancer using biological feature-based classification.* 2017. **213**(4): p. 394–399.

90. Zhang, B., et al., *Radiomic machine-learning classifiers for prognostic biomarkers of advanced nasopharyngeal carcinoma.* 2017. **403**: p. 21–27.

91. Shah, F.P. and V. Patel. *A review on feature selection and feature extraction for text classification.* in *2016 International Conference on Wireless Communications, Signal Processing and Networking (WiSPNET).* 2016. IEEE.

92. ElAlami, MEJASC, *A new matching strategy for content based image retrieval system.* 2014. **14**: p. 407–418.

93. Demel, M.A., et al., *Predictive QSAR models for polyspecific drug targets: The importance of feature selection.* 2008. **4**(2): p. 91–110.

94. Zhou, L.-T., et al., *Feature selection and classification of urinary mRNA microarray data by iterative random forest to diagnose renal fibrosis: a two-stage study.* 2017. **7**(1): p. 1–9.

95. Yousef, M., J. Allmer, and W. Khalifa, *Feature selection for microRNA target prediction comparison of one-class feature selection methodologies.* 2016.

96. Dong, C., et al., *comparison and integration of deleteriousness prediction methods for nonsynonymous SNVs in whole exome sequencing studies.* 2015. **24**(8): p. 2125–2137.

97. Pavlovic, M., et al., *DIRECTION: a machine learning framework for predicting and characterizing DNA methylation and hydroxymethylation in mammalian genomes.* 2017. **33**(19): p. 2986–2994.

98. Goh, W.W.B. and L.J.J.o.P.R. Wong, *NetProt: complex-based feature selection.* 2017. **16**(8): p. 3102–3112.

99. Saeys, Y., I. Inza, and P.J.b. Larrañaga, *A review of feature selection techniques in bioinformatics.* 2007. **23**(19): p. 2507–2517.

100. Singh, K.P., N. Basant, and S.J.A.c.a. Gupta, *Support vector machines in water quality management.* 2011. **703**(2): p. 152–162.

101. Ma, S. and J.J.B.i.b. Huang, *Penalized feature selection and classification in bioinformatics.* 2008. **9**(5): p. 392–403.

102. Lorena, L.H., et al., *Filter feature selection for one-class classification.* 2015. **80**(1): p. 227–243.

103. Kohavi, R. and G.H.J.A.i. John, *Wrappers for feature subset selection.* 1997. **97**(1–2): p. 273–324.

104. Sheng, L., et al. *Microarray classification using block diagonal linear discriminant analysis with embedded feature selection.* in *2009 IEEE International Conference on Acoustics, Speech and Signal Processing.* 2009. IEEE.

105. Guan, D., et al., *A review of ensemble learning based feature selection.* 2014. **31**(3): p. 190–198.

106. Yang, C., et al., *Clustering genes using gene expression and text literature data.* Proc IEEE Comput Syst Bioinform Conf, 2005: p. 329–40.

107. Liu, Y., et al., *Text mining biomedical literature for discovering gene-to-gene relationships: a comparative study of algorithms.* IEEE/ACM Trans Comput Biol Bioinform, 2005. **2**(1): p. 62–76.

108. Burkart, M.F., et al., *Clustering microarray-derived gene lists through implicit literature relationships.* Bioinformatics, 2007. **23**(15): p. 1995–2003.

109. Datta, S. and S. Datta, *Methods for evaluating clustering algorithms for gene expression data using a reference set of functional classes.* BMC Bioinformatics, 2006. **7**: p. 397.

110. Oyelade, J., et al., *Clustering Algorithms: Their Application to Gene Expression Data.* Bioinformatics and Biology Insights, 2016. **10**: p. 237–253.

111. Erola, P., J.L.M. Björkegren, and T. Michoel, *Model-based clustering of multi-tissue gene expression data.* Bioinformatics, 2020. **36**(6): p. 1807–1813.

112. Zhao, L. and M.J. Zaki. *Tricluster: an effective algorithm for mining coherent clusters in 3d microarray data.* in *Proceedings of the 2005 ACM SIGMOD international conference on Management of data.* 2005.

113. Chandrasekhar, T., K. Thangavel, and E.J.a.p.a. Elayaraja, *Effective clustering algorithms for gene expression data.* 2012.

114. Jiang, D., et al., *Cluster analysis for gene expression data: a survey.* 2004. **16**(11): p. 1370–1386.

115. Abu-Jamous, B. and S. Kelly, *Clust: automatic extraction of optimal co-expressed gene clusters from gene expression data.* Genome Biology, 2018. **19**(1): p. 172–172.

116. Tellaroli, P., et al., *Cross-Clustering: A Partial Clustering Algorithm with Automatic Estimation of the Number of Clusters.* PLoS One, 2016. **11**(3): p. e0152333.

117. Hand, DJ and NA. Heard, *Finding groups in gene expression data.* J Biomed Biotechnol, 2005. **2005**(2): p. 215–25.

118. Pirim, H., et al., *Clustering of High Throughput Gene Expression Data.* Computers & Operations Research, 2012. **39**(12): p. 3046–3061.

119. Teran Hidalgo, S.J., et al., *Overlapping clustering of gene expression data using penalized weighted normalized cut.* Genetic Epidemiology, 2018. **42**(8): p. 796–811.

120. Saelens, W., R. Cannoodt, and Y. Saeys, *A comprehensive evaluation of module detection methods for gene expression data.* Nature Communications, 2018. **9**(1): p. 1090–1090.

121. Futschik, M.E. and B. Carlisle, *Noise-robust soft clustering of gene expression time-course data.* J Bioinform Comput Biol, 2005. **3**(4): p. 965–88.

122. Tjaden, B., *An approach for clustering gene expression data with error information.* BMC Bioinformatics, 2006. **7**: p. 17–17.

123. Friedman, N., et al., *Using Bayesian Networks to Analyze Expression Data.* Journal of Computational Biology : A Journal of Computational Molecular Cell Biology, 2000. **7**: p. 601–20.

124. Su, C., et al., *Using Bayesian networks to discover relations between genes, environment, and disease.* BioData Mining, 2013. **6**(1): p. 6.

125. Barabási, A.L. and Z.N. Oltvai, *Network biology: understanding the cell's functional organization.* Nat Rev Genet, 2004. **5**(2): p. 101–13.

126. Blais, A. and BD. Dynlacht, *Constructing transcriptional regulatory networks.* Genes Dev, 2005. **19**(13): p. 1499–511.

127. Li, W., D. Notani, and M.G. Rosenfeld, *Enhancers as non-coding RNA transcription units: recent insights and future perspectives.* Nat Rev Genet, 2016. **17**(4): p. 207–23.

128. Lewis, M.W., S. Li, and H.L. Franco, *Transcriptional control by enhancers and enhancer RNAs.* Transcription, 2019. **10**(4–5): p. 171–186.

129. Bock, C. and T. Lengauer, *Computational epigenetics.* Bioinformatics, 2008. **24**(1): p. 1–10.

130. Landt, S.G., et al., *ChIP-seq guidelines and practices of the ENCODE and modENCODE consortia.* Genome Res, 2012. **22**(9): p. 1813–31.

131. Fan, Y., X. Wang, and Q. Peng, *Inference of Gene Regulatory Networks Using Bayesian Nonparametric Regression and Topology Information.* Comput Math Methods Med, 2017. **2017**: p. 8307530.

5 Implementation of Donor Recognition and Selection for Bioinformatics Blood Bank Application

Soobia Saeed, N.Z. Jhanjhi, and Memood Naqvi

CONTENTS

DOI: 10.1201/9781003126164-5

5.1   INTRODUCTION

Blood is the most important element of every human life, without which life cannot survive. That is why donating blood to save someone's life is considered very admirable across the globe. Blood is a perishable product that has a limited life after donation. So, to increase its life, it is compulsory to maintain its temperature. A blood bank is a place where all the blood and its products are stored. It is a kind of laboratory where all the processes related to the collection and issuance of blood take place. All the blood collected from donors is generally kept in large refrigerators to sustain the blood temperature to increase its life. Blood banks also conduct blood testing to identify major diseases in the blood to avoid issuing the patients reactive blood. An automated bioinformatics blood bank application can efficiently manage the donor registration, inventory of blood, blood screening, cross-matching, and issuance. Blood banks have in place certain criteria for the donation of blood all over the globe; hence, to donate blood a donor must fulfill the following conditions [1]:

Donor age must be between 18 and 60 years.

   i. The weight of the donor must be above 45 kgs.
  ii. The minimum hemoglobin level should be 12.5gm%.
 iii. The last donation was more than three months ago.

Blood banks do not allow blood donations if the donor possesses the following conditions:

   • If the donor has had a fever/cold in the past week
   • If the donor is taking any antibiotics or any other medications

- If the donor has any of the following problems:
  a. Cardiac
  b. Diabetes (or insulin therapy)
  c. Hypertension
  d. Epilepsy
  e. Cancer history
  f. Liver disease
  g. Chronic kidney problems
  h. Bleeding tendencies
  i. Venereal diseases etc.
- If the donor has had any serious surgery in the past six months
- If the donor has had any type of vaccination in the past 24 hours
- If the donor has fainted during a previous donation
- If the donor is using alcohol or any drugs using a needle
- If the donor is HIV positive

5.1.1 ABOUT SOFTWARE APPLICATION DEVELOPMENT

Nowadays, software applications are used worldwide due to which software application development is increasing day by day. Due to complex user requirements and newly emerging technologies, software development has become more complicated; to facilitate the user requirements and their business processes it is appropriate that each phase of the software development should be properly analyzed with a deep investigation of user requirements and business processes. To simplify the different development phases an organized method of AI, software development life cycle (SDLC), was introduced in the late 1970s and has proven to be the standard of software development and is still in practice in different forms. Figure 5.1 shows diverse strides of the frameworks advancement life cycle. In the meantime, it was understood that a product-designing procedure ought to be taken as something other than what's expected from SDLC. The programming procedure model is viewed as suitable for the improvement of a framework, every one of the periods of SDLC must be taken after. At the end of the day, procedure models particular in characterizing the arrangement of SDLC stages and every stage are key for each procedure model. Although the practices referred to above have upheld altogether ineffective frameworks improvement, one of the real reasons for disappointment, even today, is a wide gap between client prerequisites which are taken in a necessary examination period of the SDLC and the product which is inevitably conveyed to the client as the yield of the framework execution period of the SDLC. Looks into being as yet working for figuring strategies that could chop down this crevice. In the framework investigation and database outline stages, reasonable displaying is a key step responsible for translating client necessities to into reality.

5.1.2 ABOUT THE BLOOD BANK AT JPMC

The blood bank at JPMC is the third biggest blood bank in the whole country, with a screening laboratory to provide quality service to all patients. It issues more than

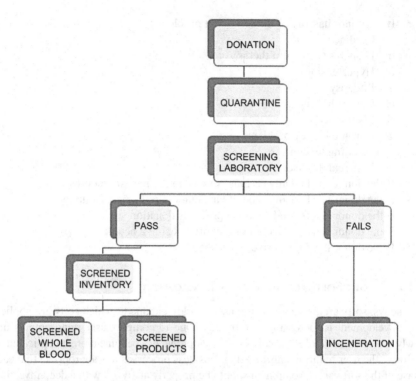

FIGURE 5.1 Process for blood collection.

60,000 pints of blood per annum. It has also an extraordinary refrigeration system to maintain and store a large number of blood bags daily. Besides screening and inventory, the bank has also facilities for the grouping of blood and crossmatching. The process at the blood bank can be divided into the following:

- Process for blood collection
- Process for blood issuance

5.1.2.1   Process for Blood Collection

The process includes the overall activity relating to the donation transactions; this includes donation entry, product extraction, and screening of blood.

5.1.2.2   Process for Blood Issuance

The process includes all the activities involved in the issuance of blood, including requisition transaction, cross-matching, and blood issuance.

5.1.3   Biometrics-AI Application

Biometrics-AI recognition systems are widely used to identify details in order to verify and distinguish evidence. Biometrics is used to manage and control access to character. In this way, the use of biometric data in the blood bank's implementation of the Taxpayer Identification Framework is a safe methodology. There are many biometric frames, such as single fingerprint recognition, face recognition, voice recognition, iris recognition, palm recognition, etc. In this company, a single framework is used for brand recognition [1].

5.1.4 WHAT IS FINGERPRINTING?

A fingerprint is a mixture of ridges and valleys on the top of a fingertip. The end points and the intersection of the ridges are called minutiae. It is widely agreed that each finger's thorough illustration is unique and does not alter in the middle of life [1].

The ridges' ends are the foci, and the forks are the place that forms part of the edge of a lonely road to two roads. Figure 5.1 shows a sample of a ridge finish and a fork. In this sample, the dark pixels show the ridges and the white pixels are related to the valleys. When unique human marking specialists examine whether two fingerprints are from the same finger, the degree of coordination between two details is one of the most essential components. Due to the similarity to human specialists' method in single fingerprints and the conservative nature of the designs, the detail-based coordination strategy is the most widely used coordination system.

Fingerprints are believed to be the most agile technique for biometric distinctiveness testing. They are safe to use, special for each individual, and do not change

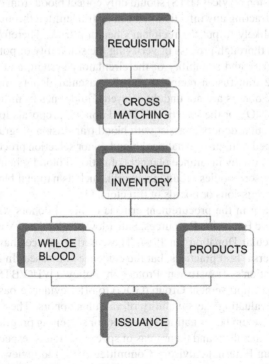

FIGURE 5.2 Process for blood issuance.

during life. Apart from these, unique brand recognition frameworks are modest, simple, and satisfactorily accurate. Single brand recognition has generally been used as part of bioinformatics applications for both legal and non-military personnel. With different and proven biometric components, single fingerprint-based biometrics is the most proven method and has the biggest piece of the pie. It is faster than other methods, and the vitality utilization of such frames is also lower [2].

5.1.5 PROBLEM STATEMENT

The aim is to design a bioinformatics blood bank application that can identify new and existing donors along with their past donation history to eliminate the risk of infected blood. It would also manage the inventory of available, reserved, and reactive blood. It also manages the complete outflow process of the bank, including requisition, cross-matching, and issuance.

5.2 LITERATURE REVIEW

This section focuses on the available literature relevant to the study. To gain maximum knowledge of relevant information, a comprehensive literature review including research papers written by different researchers, articles, white papers, and various internet blogs was undertaken.

Blood transfusion services (BTS) should only collect blood from donors who have a low risk of contracting any infection that can be transmitted through a blood transfusion and is unlikely to put the recipient's health at risk. Therefore, it is imperative to conduct a thorough process of verifying the suitability of potential donors to protect the integrity and suitability of the circulatory system, and to maintain the suitability of the transfusion recipient and the vascular donors themselves, while ensuring suitable donors are not unduly delayed. Guidelines from the World Health Organization (WHO) for the selection of blood donors: proposals for evaluating the suitability of vascular donors to assist with blood transfusion programs in countries that have developed or improved national blood donor selection processes. They may be appropriate for use by insurance planners of national blood schemes in ministries of health and advisory bodies around the world, such as national blood and transfusion services commissions or boards of directors. [3].

WHO assistance in the procurement criteria of blood donors was initially published in "Inclusive Education Resources, Safe Flow of Blood and Vascular Products, Part 1: Safe Vascular Donation" in 1994. These earlier devices have been built on the basis of universal best practices, but the evidence is focused. In 2009, the World Health Organization's Transfusion Protection Policy (WHO/BTS) assembled a Recommendation Improvement Group (GDG) to offer evidence-based guidance on the criteria for evaluating the suitability of vascular donors. The GDG has recognized the need to set up better national blood donor screening programs. The knowledge of the GDG members and their areas of specialization is respected. The WHO/BTS has set up an External Advisory Committee (ERG) to review and discuss the draft plans at different implementation stages [4].

The ERG committee consists of members of the WHO Expert Advisory Community on Blood Transfusion Medicines and experts from the WHO Collaborating Centers for Blood Transfusion Medicines, as well as representatives of national vascular transfusion programs and program specialists from each area of the World Health Organization. The review committee's role is to evaluate draft standards and make WHO recommendations on the appropriateness, applicability, and viability of advice. The inter-regional workshop on vascular donor collection and donor guidance is targeted at different countries in parts of Africa and the Eastern Mediterranean, June 2011, Nairobi, Kenya. The donor suitability criteria for a donation of blood vessels emerged in the context of the WHO Recommendation Creation Process, which includes a systematic review of current research on key problems and advice, consideration of program viability, and pricing consequences of possible new advice. During the period 1995–2015, as well as in 2016, a comprehensive analysis of written and "gray" books was undertaken to define trends. Unique programs are planned to identify systematic surveys of books and knowledge directly relevant to vascular donor selection in low- and middle-income countries. However, high-quality data promoting decisions on prospective donors' suitability to donate blood remains limited or even without consideration for certain medical conditions and risks [4, 5]. The published results were non-existent, advice is taken from international recommendations, and data and expertise for the Instruction Production Community and members in the External Evaluation Group are in human physiology, anatomy, and specialist medical therapies. In cases where there is proof that deferment criteria may be calm, a precautionary approach is recommended before good test protection is possible. The advice in this record is intended to remain in place until 2017. An overview of these recommendations is provided to explore any fresh facts, particularly on contentious issues or where the improvements used might be relevant. No trials have been undertaken to examine the occurrence of transfusion-borne diseases in previous (more than 12 months) opioid consumers. Just one paper (Musto et al.) looked at the possibility of HIV infection in infrequent consumers of drugs (once a year), but did not indicate which drugs were used. Using a statistical model, they assessed that one infusion drug use scene was associated with HIV transmission consent during the window span and concluded that it would seem appropriate to consider minimizing the HIV risk-related refuse period. In either case, it has been reported that parenteral drug use is also a major risk factor for other blood-borne infections, especially the hepatitis C virus, which has a prevalence rate of 54% in Australia. While attention was given to the decrease in HCV transmission during the time window by introducing NAT studies, the originators did not explain the difference in approach. Two studies discussed the risks of blood-borne infectious diseases in anabolic steroid consumers, with entirely different findings. Champlin et al. found a low level of needle and syringe sharing and a large decline in the frequency of HBV and HIV markers among 149 anabolic steroid clients relative to those who injected heroin and amphetamine, and thought that they could be deemed a private gathering of regard for lifestyle and infusion activities. Aitken also observed that HCV depositing decreased prevalence among 63 illicit anabolic steroid injectors compared to other drug injections, but it was found that HCV shims elucidate other unsafe activities and suggested that steroid injectors should not be removed in attempts to prevent blood-borne infections. The salmon test studied the impact of restorative stewardship on HIV prevalence in injecting drug users and observed a total involvement of 2%, separately linked to the actions of MSM and the use of psycho-stimulants. These tests indicate that infrequent usage of injected drugs, steroid use alone, and medical monitoring could minimize the risk of blood-borne illness among users of injected drugs; nevertheless, the concept of modifying the provision for postponing blood donation over the long term cannot be legitimized [5].

The research concentrates on the existing literature available which is relevant to the study. To obtain maximum knowledge of the related information, an extensive literature has been reviewed, including research papers written by different researchers, articles, white papers, and different blogs on the internet. Thus, this chapter is divided into parts according to different aspects which are considered during the information gathering for the study [6].

In their paper, Nahla Aljojo et al. have stated that a web-based blood donation management system is a management system website that enables individuals who want to donate blood to help the needy [5]. The paper proposes a web-based application form where the donor can register themselves to donate blood to needy patients. The application is divided into three segments, Donation, Requisition, and Inventory management. Donor registration is done using the Donation section. On the other hand, patients can raise requisitions using Requisition section. Inventory management is carried out by the hospitals where all the blood needs to be stored and issuance may take place.

Hassan, Otman Mohamed M. has presented the idea of a WAP-based blood bank, which allows volunteer donors to register themselves for donations to increase blood availability for requisitions. The aim is to solve the problem experienced in traditional blood donation applications [6]. The researcher has used mobile technology for developing an application which donors can easily access.

The researcher has presented an online blood donation reservation and management system (OBDRMS). This application is developed by using JSPI Servlet technology from J2EE with MySQL 5.0 as the database management system [8]. The application has been designed to overcome the current system's deficiencies and provide a centralized database solution from which patients can reserve the blood through an online application. The solution is also equipped with password securities to secure the data from unauthorized access, donation and requisition, and inventory reports including reminders of blood shortages, and workflows that have the added enhancement of the application.

The researcher has presented a paper to improve the understanding of the blood supply chain and its proper management to reduce shortage and wastage factors. The problem lies in the stochastic nature of donated and transfused blood units, which are short-lived and face exceedingly stringent consistency requirements [9]. Risk analysis and multiple regression analysis were used to determine different experimental models to improve the supply chain.

The researcher, Howe, described the challenges that occur in the inventory management and production of blood platelets. Blood platelets have a very short life, with an average life span of eight to nine days [9]. It is necessary to maintain a balanced inventory of platelets according to demand and supply. To solve this problem, a five-step procedure was used [10]. The issue is overcome using inventory management on the basis of order up to level where quantities are fixed for each day. This has eliminated the wastage percentage up to 15–20%.

Debidutt Acharya et al. have proposed a biometric solution for software application "Student Attendance System." The goal can be disintegrated into finer sub-targets: fingerprint capture and transmission, fingerprint analysis, and wireless data transfer in a server-client environment [11]. The paper aims to develop a software application through which students can mark their attendance using their thumb impression. Students enroll with the application; the system captures their thumb impression and saves the record in the database. The database was maintained on a server which is connected wirelessly with different client machines available in each class. Students need to mark their attendance using thumb scanners attached to the client machine in each class which updates the database. Attendance and other related reports are available in the system, which can be printed easily [11].

AABB has suggested some important points in the development of blood bank systems. The crucial first step in launching the program is to hire clinical champions who will make a clear argument for educating hospital leadership on the program's patient care implications, emphasizing better patient outcomes [12]. The paper discusses the planning, implementation, and maintenance as the most important phases of blood application development.

The researcher has recommended alternative policies based on a computer simulation model to obtain cost-effective blood supply management in the UK [13]. A detailed study was conducted using primary and secondary data for statistical and detailed analysis for different parts of the UK.

A local software house in Bangalore, India, has developed a software application for a local blood bank. The scale of operations had increased and they were looking for a system that could streamline their processes and improve efficiency [14]. The application's development aims to automate the whole blood banking process, including donor registration, grouping, screening, inventory management, requisition management, compatibility testing, issuance, billing, etc.

The author has presented the role of a software engineer in the development of a blood bank application. In this paper, the researcher discusses the various models used in software engineering [15]. Different design models are discussed in the paper which can be used in the blood bank management information system. The researchers have also presented and proposed the waterfall method in which the whole process is in a sequential flow and is implemented in the live environment.

The author has submitted a research paper proposing a web-based blood bank application which can be accessed globally and from where the inventory of blood can be tracked from different locations. The proposal suggests ways to overcome the issues of delay in the issuance of blood due to the communication gap between different inventory locations, resulting in unnecessary disturbance to the patients, which can be very critical. All the inventory sites are suggested to interlink so that tracking can be eased through a web-based application and patients can be dealt with in a timely manner [16].

5.2.1 DATA COLLECTIONS AND INTERVIEWS

To obtain firsthand and accurate knowledge of the processes involved in blood banks, different local blood banks are also visited and business processes are discussed in detail with different members including data entry users, inventory managers, and blood bank managers. A few of them are discussed below.

Blood bank at Indus Hospital: a state-of-the-art software application is used for internal blood bank processes. All the processes are maintained systematically where physicians generally raise requisitions on behalf of patients admitted in the hospital using a separate portal. To maintain a paperless environment no printable reports are used; however, doctors can review the data using inquiry windows for requisitions, cross-matching, and details of transfusion reaction. Cross-matching is also done using the application platform. The application solely fulfills its internal processes and has no structure for entertaining external patients [17].

Hussaini Blood Bank (Karachi): this is a private organization with numerous branches all over the city. The organization uses a custom-built application which covers all the processes of the blood bank. The application entertains all the requisitions as external patients, where a request or can get blood by donating blood in the same quantity. The bank also charges some service fees for screening and grouping donated blood. The system has no provision for any internal patient as no such requirement exists [18].

Blood Bank JPMC (Karachi): this is a government-owned blood bank situated in the Jinnah Post Medical Center managed by a local NGO. It is one of the biggest blood banks in the city having a large volume of inflow and outflow of blood. Unlike other blood banks as discussed earlier this blood bank accepts both types of patient, i.e.

    i. Internal patients (JPMC patients)
    ii. External patients (patients from other hospitals)

No charges are accepted from internal patients. External patients are required to pay service charges for requisition. However, charges can be waived if the requestor is eligible for zakat. Donors are compulsory for both types of patients. The bank has a laboratory for the screening of blood and its products. The blood bank also provides blood to other blood banks and also accepts bulk donations from other blood banks as well [19].

5.2.1.1 Blood Compatibility

Some blood groups are compatible with selected blood groups depending on the product type. It is really necessary to maintain the compatibility options as, in case of emergency, the compatible blood group can be used in transfusions of blood. Different kinds of literature were reviewed to understand the complete compatibility chart. The following is the list of cheats for each group [20][21].

Graph 5.1 shows the results of the A+ blood group which indicates the range of blood group samples as we can see the situation of datasets of the blood bank group for the treatment of patients as per the blood sample report.

Graph 5.2 shows the results of the A– blood group which indicates the range of blood group samples as we can see the situation of datasets of the blood bank group for the treatment of patients as per the blood sample report.

Graph 5.3 shows the results of the B+ blood group which indicates the range of blood group samples as we can see the situation of datasets of the blood bank group for the treatment of patients as per the blood sample report.

Graph 5.4 shows the results of the B– blood group which indicates the range of blood group samples as we can see the situation of datasets of the blood bank group for the treatment of patients as per the blood sample report.

TABLE 5.1
Blood Compatibility Chart for A+ Blood Group [21][22]

	A+	A–	B+	B–	AB+	AB–	O+	O–
Whole blood	√	√	X	X	X	X	X	X
Platelets	√	√	√	√	√	√	√	√
WPC	√	√	X	X	X	X	√	√
MU	√	√	X	X	X	X	X	X
Plasma	0	0	0	0	0	0	0	0
CP	√	√	√	√	√	√	√	√
FFP	√	√	X	X	√	√	X	X
PC	√	√	X	X	X	X	√	√

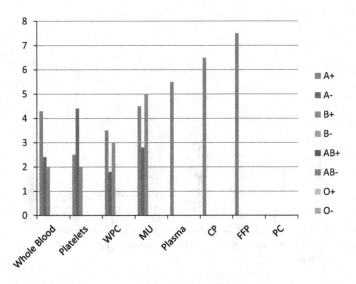

GRAPH 5.1   Blood Compatibility Chart for A+ Blood Group

Graph 5.5 shows the results of the AB– blood group which indicates the range of blood group samples as we can see the situation of datasets of the blood bank group for the treatment of patients as per the blood sample report.

Graph 5.6 shows the results of the O+ blood group which indicates the range of blood group samples as we can see the situation of datasets of the blood bank group for the treatment of patients as per the blood sample report.

Graph 5.7 shows the results of the O– blood group which indicates the range of blood group samples as we can see the situation of datasets of the blood bank group for the treatment of patients as per the blood sample report.

Graph and Table show the statistical range of blood groups of human beings in terms of their functionalities.

TABLE 5.2

Blood Compatibility Chart for A– Blood Group [21]

Products	A+	A–	B+	B–	AB+	AB–	O+	O–
Whole blood	X	X	√	√	X	X	X	X
Platelets	√	√	√	√	√	√	√	√
WPC	X	X	√	√	X	X	√	√
MU	X	X	√	√	X	X	X	X
Plasma	0	0	0	0	0	0	0	0
CP	√	√	√	√	√	√	√	√
FFP	X	X	√	√	√	√	X	X
PC	X	X	√	√	X	X	√	√

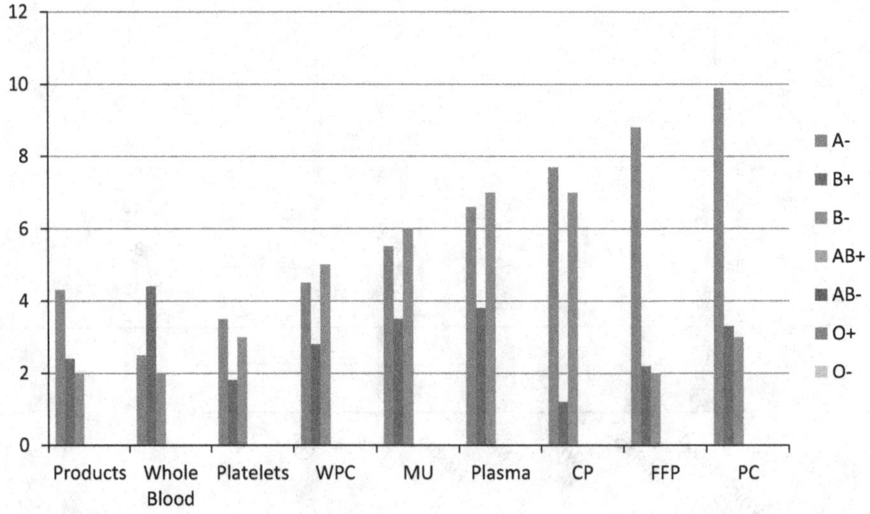

GRAPH 5.2   Blood Compatibility Chart for A– Blood Group

TABLE 5.3

Blood Compatibility Chart for B+ Blood Group [21][22]

Products	A+	A–	B+	B–	AB+	AB–	O+	O–
Whole blood	X	X	X	√	X	X	X	X
Platelets	√	√	√	√	√	√	√	√
WPC	X	X	X	√	X	X	X	√
MU	X	X	X	√	X	X	X	X
Plasma	0	0	0	0	0	0	0	0
CP	√	√	√	√	√	√	√	√
FFP	X	X	X	√	√	√	X	X
PC	X	X	X	√	X	X	X	√

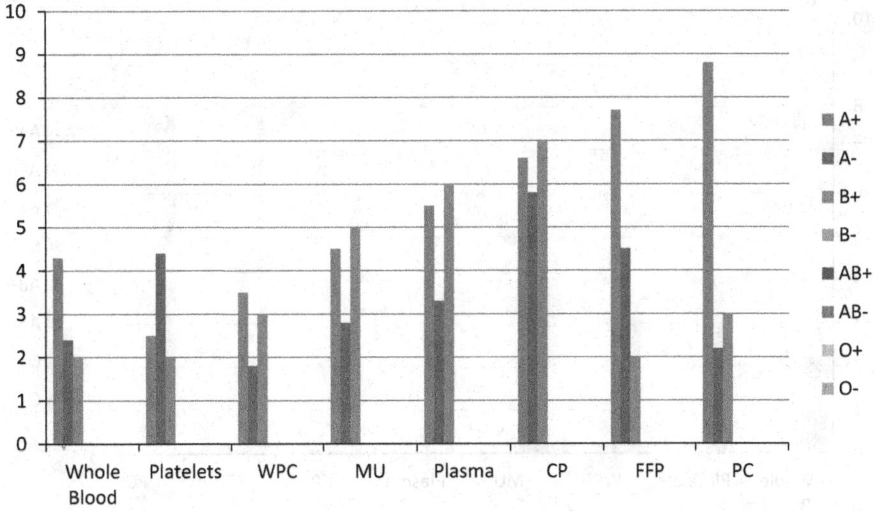

GRAPH 5.3   Blood Compatibility Chart for B+ Blood Group

5.3   METHODOLOGY

5.3.1   Bioinformatics Blood Bank Application Framework

Using manual data entry in organizations like blood banks, where the flow of trans-actions is very high, can cause many data management abnormalities. It takes too many resources in terms of cost and human effort to record routine transactions and manage inventory daily. In blood banks, it is really important to keep an eye on the inventory so that blood can be utilized properly to decrease blood expiry chances. Donors cannot be identified properly and therefore donors are generally regis-tered every time, even if they have already registered. The identification of donors with reactive blood on their last donation may also not be known, and therefore

TABLE 5.4

Blood Compatibility Chart for B– Blood Group [21][22]

Products	A+	A–	B+	B–	AB+	AB–	O+	O–
Whole blood	X	X	X	X	√	√	X	X
Platelets	√	√	√	√	√	√	√	√
WPC	√	√	√	√	√	√	√	√
MU	X	X	X	X	√	√	X	X
Plasma	0	0	0	0	0	0	0	0
CP	√	√	√	√	√	√	√	√
FFP	X	X	X	X	√	√	X	X
PC	√	√	√	√	√	√	√	√

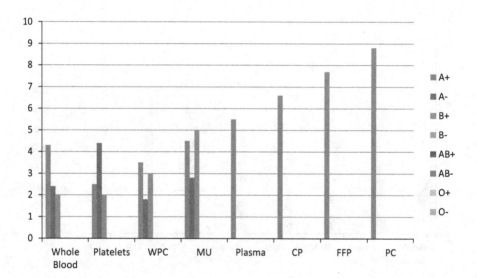

GRAPH 5.4   Blood Compatibility Chart for B– Blood Group

the chances of getting reactive blood are also increasing. Screening is also done manually causing detailed paperwork to be done manually to bifurcate the inventory of screened and unscreened blood. Requisitions are also managed manually which again increases the human efforts on one hand and the reporting complications on the other. Issuance is also done manually and therefore inventory cannot be calculated on a real-time basis. To avoid these unnecessary irritations and transaction burdens on users and top-level management, a blood bank application has been designed and implemented to reduce the resources required in terms of cost and human efforts and to efficiently manage the donor registrations and inventory management.

The system is designed to replace all the manual data entries with an automated software application that can identify and select the donors effectively using thumb

TABLE 5.5

Blood Compatibility Chart for AB– Blood Group [21][22]

Products	A+	A–	B+	B–	AB+	AB–	O+	O–
Whole blood	X	X	X	X	X	√	X	X
Platelets	√	√	√	√	√	√	√	√
WPC	X	√	X	√	X	√	X	√
MU	X	X	X	X	√	√	X	X
Plasma	0	0	0	0	0	0	0	0
CP	√	√	√	√	√	√	√	√
FFP	X	X	X	X	√	√	X	X
PC	X	√	X	√	X	√	X	√

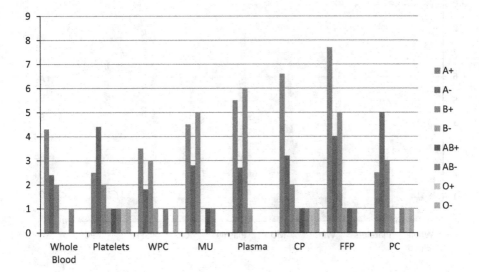

GRAPH 5.5 Blood Compatibility Chart for AB– Blood Group

impressions. Blood bags are marked with barcode stickers to avoid the manual writing of donation numbers on blood bags and sample tubes to properly and effectively track the blood bag details. Screening results are also entered through the application platform where a user can mark each blood bag's screening results. The product extraction process is also done using separate from where the user can extract products from a whole blood bag. The system manages the inventory of blood bags according to the blood groups and products. The user can view the inventory of each product of any group at any time. The system also manages the expiry date of all the blood bags according to the products.

Further to the above, the system also allows users to record requisitions in the system. Requisition slips and barcode stickers are also printed here for the proper

TABLE 5.6
Blood Compatibility Chart

Products	A+	A–	B+	B–	AB+	AB–	O+	O–
Whole blood	X	X	X	X	X	X	√	√
Platelets	√	√	√	√	√	√	√	√
WPC	X	X	X	X	X	X	√	√
MU	X	X	X	X	X	X	√	√
Plasma	0	0	0	0	0	0	0	0
CP	√	√	√	√	√	√	√	√
FFP	√	√	√	√	√	√	√	√
PC	X	X	X	X	X	X	√	√

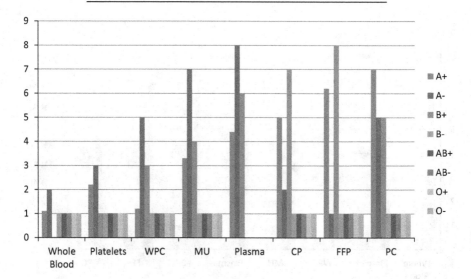

GRAPH 5.6   Blood Compatibility Chart for O+ Blood Group

identification of requisition slips with a sample tube as provided for blood sampling and cross-matching. The system also manages cross-matching results and reserves blood bags for particular requisitions so that a blood bag cannot available for any other requisition. The system also manages the cancellation of the reservation after three days if blood bag is not issued to the particular requisition. The system also has an option to record the blood bag return where the user can enter the returned bags. Moreover, the user can also incinerate the reactive or expired blood bags available in the system upon which system updates the inventory of blood bags accordingly.

5.3.2 System Analysis

The most important phase of the SDLC consists of different sub-phases; this phase is to obtain maximum information regarding the processes and workflow of the entire

TABLE 5.7

Blood Compatibility Chart for O– Blood Group [21][22]

Products	A+	A–	B+	B–	AB+	AB–	O+	O–
Whole blood	X	X	X	X	X	X	X	√
Platelets	√	√	√	√	√	√	√	√
WPC	X	X	X	X	X	X	X	√
MU	X	X	X	X	X	X	√	√
Plasma	0	0	0	0	0	0	0	0
CP	√	√	√	√	√	√	√	√
FFP	√	√	√	√	√	√	√	√
PC	X	X	X	X	X	X	X	√

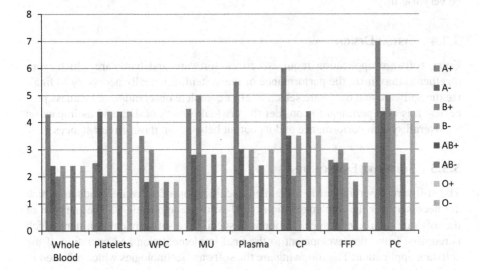

GRAPH 5.7   Blood Compatibility Chart for O– Blood Group

business. The information is generally gathered from different means; a few of them are discussed here.

5.3.3   GATHERING INFORMATION

This is an essential phase which plays a prominent role in gathering maximum information from a different level of users. This phase helps in the collection of basic to complex processes which are needed to accommodate the system.

- The information is gathered from the following user levels.
- Highest-level administration

- Moderate-level administration
- Low-level administration

5.3.3.1 Observation

It is also a part of system analysis in which data is generally gathered after observing the processes by visiting the premises and all of its live processes at different occasions like at the time of extreme flow of transactions and at times where transaction flow is low.

5.3.3.2 Record Review

It is also important in the system analysis to collect data from history. It provides a clear picture of what data is generally kept by the firm and what are the reporting requirements which need to be addressed in the proposed solution and also allows the identification of loopholes and missing information which must be kept and can be very useful.

5.3.4 SYSTEM DESIGN

Every software application requires certain software and hardware which work together to maximize the performance of the system. It is really necessary to finalize the software and hardware selection after complete observation of business processes. It is also pertinent to consider the gradual growth of data and its impact on the overall system performance and its output behavior in those circumstances.

5.3.5 SOFTWARE ENVIRONMENT

The software environment is a detailed specification of software products which are necessary for the implementation of AI-software applications. The purpose of the software requirement specification is to specify the software environment which is required from the development to the final implementation and operation of the software application. The following are the software technologies which are used in the development of the application.

5.3.6 AI-SOFTWARE APPLICATION PLATFORM

Microsoft Visual Studio 2010 is used for the development of a software application and the UI interfaces of the application. It contains versatile features that enable developers to work easily and efficiently while working with different programming languages and designing UI. VB programming language is used for writing the code of the application, which uses signs all over the globe.

5.3.7 DATABASE MANAGEMENT SYSTEM (DBMS)

Microsoft SQL Server 2008R2 is used as the database management system for the application. Microsoft Structured Query Language (SQL) is a widely used database management system which provides very easy options/steps for reading and writing data in relational databases. Codes/queries are written in the English language which can easily be understood, and users can easily get data from the database.

5.3.8 REPORTING ENVIRONMENT

5.3.8.1 Crystal Reports

Crystal Reports XI is used for designing the reports that must be printed from the application. The following reports are designed and implemented using Crystal Reports.

- Barcode stickers for blood bags
- Barcode stickers for blood sample tube
- Barcode stickers for a patient's blood sample tube
- Patients' requisition slip printing reports
- Blood screening result reports for donors
- Blood cross-matching report for patients

5.3.8.2 SQL Server Reporting Services (SSRS)

SQL Server Reporting Services are used to print reports for managerial levels. SSRS reports are published on the web and can be viewed easily using a web browser. The following reports are built on SSRS.

1. Donation register
2. Requisition register
3. Inventory register
4. Shift wise collection and disbursement of blood
5. Overall shift information

5.3.9 HARDWARE AND SOFTWARE ENVIRONMENT

The hardware environment includes all the physical hardware and software which is required during the pre-implementation and post-implementation of the software application.

5.3.10 SYSTEM DEVELOPMENT

The most important and critical part of any software application is its development phase where all the requirement analysis and system designs that are finalized in earlier phases are taken into consideration to develop the database and software application. Both the AI-software application and database development generally travel parallel to each other in the development process, and change in one process can impact the other as well.

TABLE 5.8
List of Hardware and Software Environments

	Server	Client
Processor	Core i3 or above	Minimum dual core
RAM	4 GB or above	Minimum 1 GB
HDD	1 TB	Minimum 40 GB
Keyboard	104 keys	104 keys
Mouse	Any optical mouse	Any optical mouse
Display	15" digital color monitor	15" digital color monitor
Fingerprint scanner	N/A	Motorola Scanner LS9208i
Barcode sticker printer	N/A	Zebra Gk420T Barcode
POS receipt printer	N/A	Birch Receipt Printer BP003u
Database management studio	SQL Server 2008R2	N/A
SQL Server Reporting Services	SQL Server 2008R2	N/A
Crystal Reports Runtime	CRRedist2008_x64	CRRedist2008_x84

5.3.11 DATABASE DESIGN

On the basis of the framework and necessity testing, a database schema has been developed that includes all the important procedures of the blood donation center. Each material is meticulously planned with its connections. The suggested materials and attributes required for the blood donation center database are followed as needed. Frame testing is the stage prior to the implementation of a framework wherein the tire is error free and all necessary adjustments are made. Tire has been tested with test information and vital tire repairs have been made. All of the reports have been verified and confirmed by the customer. The framework was exceptionally easy to use with online customer assistance where needed.

5.3.12 ALPHA TESTING

A client did this on the designer's site. The product is used as part of a unique environment where the engineer "looks over the shoulder" of the customer and records errors and usage issues. Alpha tests are performed in a controlled setting.

5.3.13 BETA TESTING

This was directed to one or more customer destinations by the end customer of the product. Unlike alpha testing, the designer often doesn't display. In this sense, a pilot test is the "direct" use of the product in an area which the engineer cannot control. The customer records all issues encountered during beta testing and reports them to the designer at regular intervals. As a result of issues reported in the middle of the beta, our programming professionals make adjustments and then prepare the product article to reach the entire customer base.

5.3.14 Test Deliverables

The following records are required other than the test arrangement:

- Unit test report for every unit
- Experiment detail for framework testing
- The report for framework testing
- Error report

5.4 RESULT AND DISCUSSION

5.4.1 Relationships

A relationship is the association between different entities. There are a few certain connections between the different entities. In fact, at whatever point the quality of one element sort alludes to another element sort, some relationship exists. The cardinality proportion for a relationship determines the most extreme number of relationship examples that a substance can take an interest in. For instance, Blood Groups: Person Master is of cardinality proportion 1:N, implying that every blood group can be identified with (that is, PersonId) any number of persons, yet an individual can be identified with only a blood group ID. This implies for this specific relationship, a specific blood group ID substance can be identified with any number of person IDs (N shows there is no largest number). Then again, an individual can be identified with at most one blood group ID. The conceivable cardinality proportions for twofold relationship types are 1:1, 1:N, N:1, and M:N.

1. One-to-One (1:1)

In this type of relationship, one entity is connected with one other entity and vice versa. If, for instance, in our blood group relationship, we specified that one individual has one blood group, then the individual/blood group relationship would be balanced.

2. Many-to-One (M:1)

In a many-to-one relationship, we can say that numerous persons are connected with one blood group and one blood group is connected with numerous persons.

3. One-to-Many (1:M)

According to the rationale of a one-to-numerous relationship, we can say that a man is connected with numerous gifts and a gift is connected with one individual. It is entirely clear that, on the off chance that we characterize a relationship as 1:M (or M:1), then we ought to be clear about which element is 1 and which is M.

5.4.2 Normalization

Normalization may be seen as improving a coherent model by erasing/minimizing knowledge peculiarities and redundancies. This technique involves the arrangement of utilitarian conditions (and different conditions, depending on what sort of system the social outline needs to be standardized) as its details. In order to rid the database of peculiarities and redundancies, systems must be structured. The standardization method involves part of the table in two or three tables of disintegration). After the tables are broken apart (a process called decay), they can be combined with a "connection" operation. Three deteriorations will minimize the standardization problems in our diagrams, as discussed below.

5.4.3   CODE DESIGN

The rationale behind the code is to promote evidence and recover recognizable records. Symbol is a necessary accumulation of photographs arranged to provide extraordinary, visible proof of an element or property. In order to obtain outstanding recognition, there must be only one place where the known entity or feature can be inserted in the code; on the other hand, there must be a place in the code for anything to be illuminated. This unrelated aspect should be used in every coding system.

- Perfect human use and computer performance.
- The length of the code varies from one character in length to five characters in length.
- No The structure of the code is exceptional.
- Guarantee that the discernible appreciation of a meaningful code can be closely related to a particular feature or attribute.
- The structure of the code is extensible provided the changing arrangement of the elements and their properties.
- The code is lightweight and short for recording, communications, dispatch, and power efficiency.
- The icons are simple enough that the user can recognize them without much effort.
- The icons are also adaptive, that is, they are difficult to adjust to represent complex shifts in state, red contour, and coded material relations.
- Added codes can also be stored easily to report a pre-defined request from the company.
- Icons are reliable and do not require subsequent analysis during much of the customer's productivity.
- Icons are relevant, too.
- It is also workable, that is to say, it is adequate for the present and suspicious knowledge that predisposes all humans.

5.5   CONCLUSION

5.5.1   IMPLEMENTATION AND EVALUATION

During the product testing phase, all programming modules are fully tested for errors and accurate results through a bioinformatics application method. The framework created is very easy to understand and the client's full documentation is provided as assistance when necessary. Usually the usage phase ends with a formal test that includes each part. The entire framework is produced using .NET technologies and an SQL server as backend. Microsoft SQL Server Reporting Services is implemented so that management can view detailed reports of all operations. Crystal Reports are used as printable reports that must be created within the application interface to provide an easy-to-use interface for users. MSSQL is the backend tool where the database is located. Hence, the entire framework diagram is easy to use and basic implementation is completely simple.

5.5.2 Results

The application has been duly completed and implemented in the blood bank and currently serves all the tasks related to both the donation and requisition processes. It has eliminated the manual paperwork, and all the transactions are generated using an automated system. By introducing a fingerprint recognition system all the donors are properly managed without any duplication. Screening process entries are also managed using the application which helps in the identification of donors who have had any reaction found on their last donation. Requisitions are also managed using the same system where all the requisitions are regularly generated for all shifts. Cross-matching transactions are also managed through the system. The issuance of blood and products is also managed through the system. The whole transaction history, including donation transactions, requisitions transactions, and inventory-related transactions, is easily accessible from reports deployed for both management and staff.

5.5.3 Limitations

Due to the lack of resources in terms of both time and cost some features in this application are compromises that can add more strength to its functionality.

1. Screening results need to be entered manually by the staff; this can be automated by the integration of a screening machine which can send the screening results directly to the application.
2. The fingerprint recognition process can be made more accurate by replacing SDK with image processing code which can apply different filters to match the thumb impression more effectively.
3. SMS generates for life-saving contact numbers and to all regular donors upon the shortage of blood.

5.5.4 Conclusion

This bioinformatics blood bank application gives an adequate opportunity to plan, code, test, and execute the application. This has offered entering into the routine of different software to implement the proposed application with engineering standards and database management ideas like keeping up respectability and consistency of information. Further, it would be beneficial to learn more about MSSQL, SSRS, Crystal Reports, Microsoft Visual Studio, and .NET to understand this bioinformatics blood bank application management system through using this technical expertise.

5.5.5 EXTENSIBILITY

Alternate elements, which the blood bank administrations give, can likewise be consolidated into this blood bank. The encryption principles can likewise be utilized to make the exchanges more secure. The current application covers the processes related to the blood only. It can be enhanced to manage the inventory of supplies, including bleeding kits, screening kits utilized for the testing of blood, etc.

BIBLIOGRAPHY

1. Alter HJ, Klein HG (2019). The hazards of blood transfusion in historical perspective. *ASH 50th Anniversary Reviews*, 7:117. doi:10.1182/blood-2008-07-077370.
2. Young C, Chawla A, Berardi V, Padbury J, Skowron G, Peter J (2012). Krause the Babesia Testing Investigational Containment Study Group. 27 March. Preventing transfusion-transmitted babesiosis: preliminary experience of the first laboratory-based blood donor screening program. Wiley online library. https://doi.org/10.1111/j.1537-2995.2012.03612.x
3. Reiss RF (2015). Blood donor well-being: a primary responsibility of blood collection agencies. *Annals of Clinical & Laboratory Science*, 1(1):3–7. Global Database on Blood Safety.
4. Eder A. et al. (2014). Selection criteria to protect the blood donor in North America and Europe: past (dogma), present (evidence), and future (hemovigilance). *Transfusion Medicine Reviews*, 23(3):205–220.
5. Moreno J (2010). "Creeping precautionism" and the blood supply. *Transfusion*, 43:840–842.
6. Boulton F (2014). Evidence-based criteria for the care and selection of blood donors, with some comments on the relationship to blood supply and emphasis on the management of donation-induced iron depletion. *Transfusion Medicine*, 18:13–27.
7. Hisham S, Al-Madani A, Al-Amri A, Al-Ghamdi A, Bashamakh B, Aljojo N (2010). Online blood donation reservation and management system In Jeddah. *Life Science Journal*, 2014; 11(8):60–65.
8. Hassan O (2010). *Mobile Blood Donation Application*. University Utara Malaysia.
9. Guðbjörnsdóttir E (2015). *Blood Bank Inventory Management Analysis*. School of Science and Engineering at Reykjavík University.
10. Vannier E, Krause PJ (2009). Update on babesiosis. *Interdisciplinary Perspectives on Infectious Diseases*.
11. Acharya D, Mishra AK (2010). *Wireless Fingerprint-Based Student Attendance System*. National Institute of Technology Rourkela.
12. Young C, Krause PJ (2009). The problem of transfusion-transmitted babesiosis. *Transfusion*, 49:2548–50.
13. Blood Bank Management System. Retrieved December 13, 2015, from http://www.jagriti.co.in/solutions/healthcare/blood-bank-management-application.
14. Kulshreshtha M (2010). Role of software engineering in blood bank management information system. *IOSR Journal of Engineering*, (Dec. 2012); 2(12):01–04.

15. Topalli D (2013). *A Database Design Methodology for Complex Systems.* The Graduate School of Natural and Applied Sciences of Atilim University.
16. Elmasri R, Navathe SB (2014). *Fundamentals of Database Systems.*
17. Thalheim B (2013). *Entity-Relationship Modeling: Foundations of Database Technology.* https://books.google.com.pk. 2013.
18. Teorey TJ, Lightstone SS, Nadeau T, Jagadish HV (2011). *Database Modelling and Design: Logical Design.*
19. Bagui S, Earp R (2011). *Database Design Using Entity-Relationship Diagrams.*
20. Lockwood Lyon. (2010). Quality and Database Design. http://www.databasejournal.com/features/db2/article.php/3888026/Quality-and-Database-Design.htm.
21. Howard J. Seven Deadly Sins of Database Design. http://edn.embarcadero.com/article/40466.
22. Mullins CS (2012). Avoiding Database Design Traps to Ensure Usability. http://www.dbta.com/Columns/DBA-Corner/Avoiding-Database-Design-Traps-to-Ensure-Usability-86581.aspx.

APPENDIX

APPLICATION PROCESS FLOW

The main application process flow starts from the point where the user logs in the system. The following are the steps for the complete process flow:

a. *User Login*

The user shall require a unique user id and password provided by his/her management according to his/her role in the blood bank. Each user shall be able to access forms according to roles and rights as permitted by the system administrator. The steps for the user log are:

- The user enters his id and password in the specified fields and clicks on the login button.
- Upon successfully clicking on the login button the system shall open the main interface form of the application from where the user can navigate to a different area of the application.

FIGURE 5.3 User login form.

FIGURE 5.4   Main form of application.

b. *Donation Process*

- The donor puts his thumb on the fingerprint scanner.
- The user clicks on the "Capture" button to save a temporary image of the thumb impression.
- User clicks on the "Search" button upon which the system shall match the captured thumb impression image within the existing records in the main database.
- Upon successful matching of the thumb impression within the database, the system shall verify the last donation status of the donor and categorize the donor in the following categories.
- If the matching of the thumb impression within the database is unsuccessful, the system shall ask the user to register the donor first.
- The user shall register the donor by defining all the donor details which shall be saved within the database along with the thumb impression.

TABLE 5.9
Donor Categories

S. No.	Category	Blood Screening Status	Acceptance for Donation
1	Active donors	Non-reactive	YES
2	Gray listed donors	Indeterminate	YES
3	Black listed donors	Reactive	NO

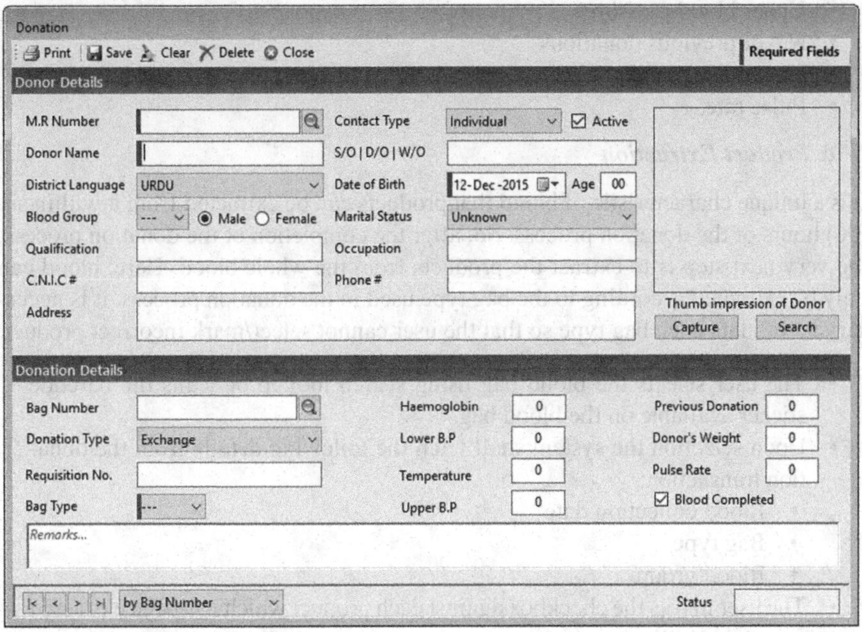

FIGURE 5.5 Donation entry form.

STEP 0	START	STEP 12	ELSE IF Match = False
STEP 1	LOAD Form Donation	STEP 13	SHOW Message "Donor not found, Create new record"
STEP 2	OPEN Connection Fingerprint Scanner	STEP 14	READ fields Donor detail record
STEP 3	CAPTURE Thumb Impression Image	STEP 15	READ fields Donation details data
STEP 4	STORE Thumb image in temp Directory	STEP 16	END LOOP
STEP 5	OPEN Database Connection	STEP 17	CLASSIFY Donor Type
STEP 6	LOOP Search Thumb image in Database	STEP 18	GET Next Donation Number
STEP 7	LOAD Table Donation Master	STEP 19	SAVE Donation record
STEP 8	MATCH Image with table records	STEP 20	SHOW Report Donation Sticker
STEP 9	IF Match = True	STEP 21	PRINT Report Donation stickers
STEP 10	SHOW Donor Details on Form Donation	STEP 22	CLOSE Database Connection
STEP 11	READ fields Donation details data	STEP 23	END

FIGURE 5.6 Donation identification and registration algorithm.

c. *Donation Details*

If the donor lies within the categories of Active or Gray Listed donors, the user shall continue to enter the donation details, including the current medical condition of the donor. The following details shall be recorded in the donation details section:

- Donation type
- Bag type
- Hemoglobin
- Lower blood pressure
- Temperature

- Upper blood pressure
- No. of previous donations
- Donors weight
- Pulse rate

d. *Product Extraction*

It is a unique characteristic of blood that products can be extracted from it within six (06) hours of the donation process. So, after the completion of the donation process, the very next step is to extract the products from the whole blood. Here, blood can only be extracted according to the bag type used in the donation process. It is necessary to validate each bag type so that the user cannot select/mark incorrect product.

- The user selects the blood bag using search lookup or scans the barcode sticker available on the blood bag.
- Upon selection the system shall fetch the following details from the donation transaction:
  - Blood collection date
  - Bag type
  - Blood group
- The user marks the checkbox against each product which has been extracted from whole blood.
- Upon clicking on the Print/Save button the system shall validate the bag type and blood products; upon successful validation the system shall print a barcode sticker for each product.

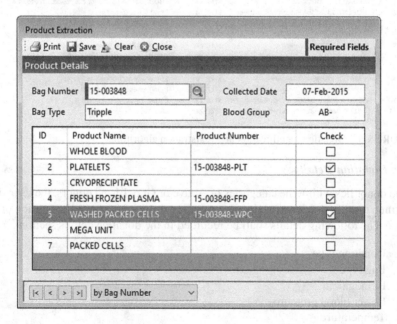

FIGURE 5.7   Product extraction form.

e. *Blood Screening*

It is a worldwide standard to screen each blood bag after donation to identify any hidden disease which could be transferred to the patient during the transfusion process. For this purpose, a separate form is designed from where user screens a bag after marking an identification or marker on it.

- The user selects the blood bag using search lookup or scans the barcode sticker available on the blood bag.
- The system validates the bag number and reflects the blood group marked at the time of donation.
- The user enters cut off values and test values against each test performed against the blood bag and based on these values mark the status of the test as:
  - Non-reactive
  - Indeterminant
  - Reactive
- If any of the tests are identified as "Reactive" or "Indeterminant" the bag status is marked as reactive after which it cannot be available for reservation or issuance to any requisition.

f. *Mass Collection*

Due to the extensive outflow of blood, it is a general process of the blood bank to receive a mass donation from other hospitals and blood banks where blood or

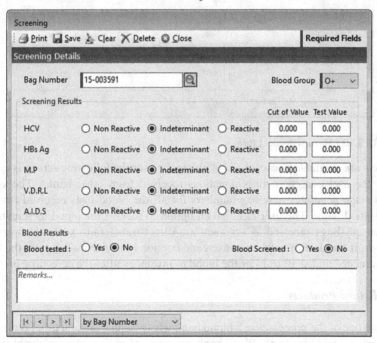

FIGURE 5.8   Screening form.

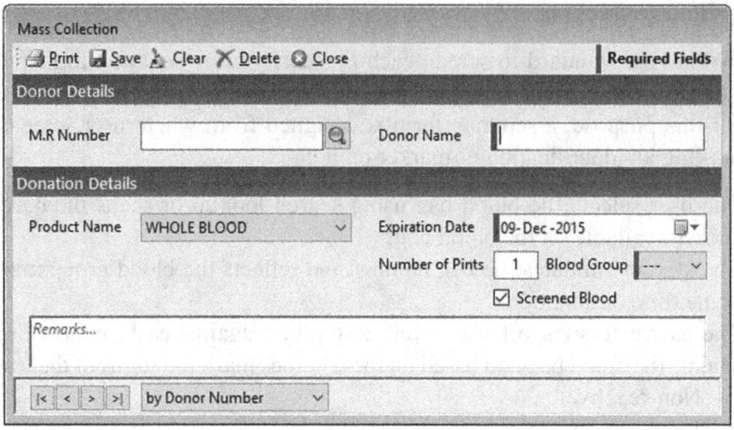

FIGURE 5.9 Mass collection form.

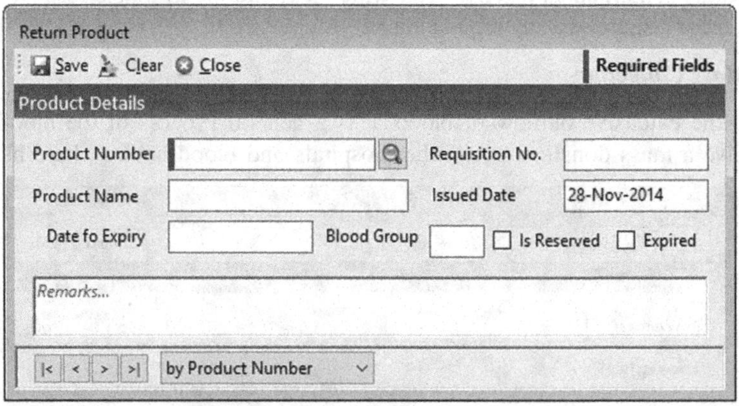

FIGURE 5.10 Return product form.

products are available in excess. To facilitate the user the following screen is created where the user can enter mass donations received from other banks. The system at this stage generates new bag numbers for all the blood bags received from the donor blood bank. The user has also an option to receive the blood as screened or unscreened. Bags received as screened are direct transfers to screened inventory; however, if the user opts for the unscreened option the system shall treat the blood as unscreened and shall transfer the blood to inventory after the screening process.

g. *Return Products*

It is also one of the processes in a blood bank where requestor sometimes returns the blood bag to the blood bank. Blood bank personnel after inspection of the blood bag receive the blood accordingly.

If the condition of the blood bag is not satisfactory, then the user marks the blood bag as expired. On the other hand, if the condition of the blood bag is found satisfactory then the user after confirmation from the requestor marks it as reserved. If the blood bag is not marked as reserved then it will be sent to unarranged inventory where it can be reserved or issued to any other bag.

h. **Requisition Process**

The very first step of the requisition process is the requisition entry form where the user enters the details of the requestor and requirement of blood or products. The process starts when the requestor comes to the requisition desk and presents the blood acquisition slip to the blood bank representative. The blood bank representative then records the following information of the patient.

i. *Patient's information*
   - Patient's name
   - Father's/husband's name

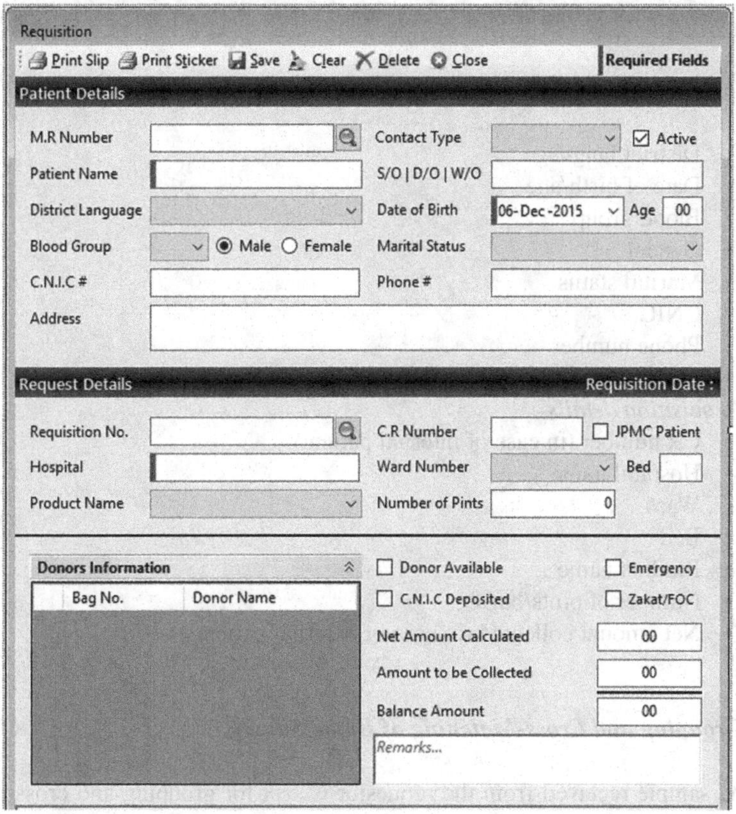

FIGURE 5.11 Requisition entry form.

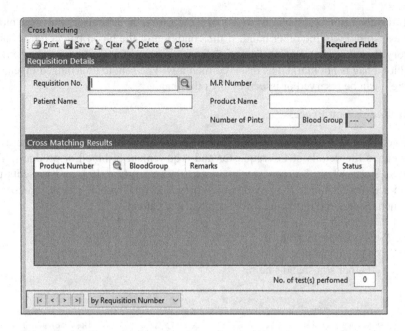

FIGURE 5.12   Cross-matching form.

- District language
- Date of birth/age
- Blood group
- Gender
- Marital status
- CNIC
- Phone number
- Address

ii. *Requisition details*
- CR number (in case of internal patient)
- Hospital name
- Ward
- Bed
- Product name
- Number of pints/bags
- Net amount collected (in case of external patients)

i. *Grouping and Cross-Matching of Blood Sample*

A blood sample received from the requestor is sent for grouping and cross-matching with available screened blood and blood that are cross-matched with each other than reserves for the requestor. The user can cross-match the blood with unarranged

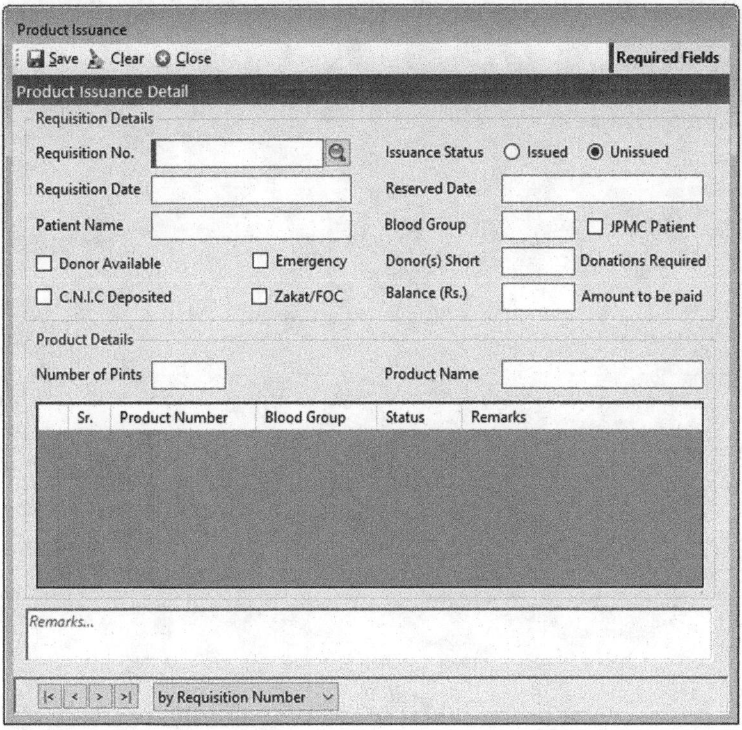

Product Issuance

Save Clear Close   **Required Fields**

Product Issuance Detail

Requisition Details

Requisition No. [   ] 🔍   Issuance Status ○ Issued ● Unissued

Requisition Date [   ]   Reserved Date [   ]

Patient Name [   ]   Blood Group [   ] ☐ JPMC Patient

☐ Donor Available   ☐ Emergency   Donor(s) Short [   ] Donations Required

☐ C.N.I.C Deposited   ☐ Zakat/FOC   Balance (Rs.) [   ] Amount to be paid

Product Details

Number of Pints [   ]   Product Name [   ]

Sr.	Product Number	Blood Group	Status	Remarks

Remarks...

|< < > >| by Requisition Number ▾

FIGURE 5.13   Product issuance form.

blood only. The system only allows the blood of the same group or groups that are compatible with the blood group of the requestor. The system does not allow cross-matching of blood that is already reserved for any other requisition. The user scans the requisition number available on the sample tube or selects from search lookup upon which the system populates the requestor details. The user scans or selects the blood bag number available on the blood bag which matches the required blood. The system validates the selected blood bag and marks it as arranged for the selected requisition number.

j. *Requisition Fulfilled*

The last requisition process is the issuance of the blood bag to the requestor. The requestor approaches the issuance desk and presents the requisition slip to the user. The user scans or selects the requisition slip to track the requisition number. The system displays the information of requisition and cross-matching along with blood bags reserved for the requisition. The pints/bags are then collected from the refrigerators containing reserved blood bags and they are issued to the requestor.

6 Deep Learning Techniques for miRNA Sequence Analysis

Saswati Mahapatra, Jayashankar Das, David Correa Martins-Jr, and Tripti Swarnkar

CONTENTS

6.1 Introduction 139
    6.1.1 Biogenesis of miRNA 140
    6.1.2 Biology behind miRNA-Target (mRNA) Interactions..................... 141
6.2 miRNA Sequence Analysis 142
6.3 Deep Learning: Conceptual Overview... 143
    6.3.1 Deep Neural Networks (DNNs).. 143
    6.3.2 Convolutional Neural Networks (CNNs)....................................... 144
    6.3.3 Autoencoders 144
    6.3.4 Recurrent Neural Networks (RNNs) 145
    6.3.5 Long Short-Term Memory (LSTM) 145
6.4 Deep Learning: Applications for Pre-miRNA Identification 145
6.5 Deep Learning: Applications for miRNA Target Prediction 147
6.6 Critical Observations and Future Directions... 148
6.7 Conclusion 149
Bibliography 149

6.1 INTRODUCTION

Ribonucleic acid (RNA) is a single-strand polymeric molecule essential in various biological processes. The building block of RNA is a chain of arbitrarily arranged nucleotides: A, U, G, C, abbreviations of adenine, uracil, guanine, and cytosine respectively. Most cellular organisms use messenger RNA (mRNA) as a carrier of genetic information that directs the synthesis of a specific protein. Micro RNAs belong to a novel category of non-coding RNAs (ncRNAs), consisting of an average of 22 nucleotides [1]. They play a crucial role in regulating gene expression at post-transcriptional level. Micro RNAs bind to partially complementary 3' UTR of target mRNA and play a significant role in the degradation of a transcript or inhibition of translation [2]. As a result, the translation of mRNA to protein is blocked.

Study reveals that miRNAs play a significant role in a plethora of regulatory events essential for both normal and disease conditions in the cell [5, 6]. The aberrant expression of miRNA facilitates many human diseases including neurodegenerative disease (Alzheimer's disease) and cancer [3,4]. Thus the identification of miRNAs as well as their targets (miRNA–mRNA interaction) is vital in understanding the complex miRNA regulating networks and exploring their therapeutic effect in disease.

6.1.1 Biogenesis of miRNA

A mature miRNA is generated through different steps of biogenesis. First, the miRNA gene is transcribed to produce primary transcripts of miRNA (pri-miRNA). Further, primary miRNAs are processed to form ~80 base pairs (bp) long, hairpin-looped precursor miRNAs (pre-miRNA). Pre-miRNAs are then further transported into the cytoplasm to undergo further processing and produce mature miRNAs having 20~23 bp [7]. Because of their short length, it is troublesome to directly identify a mature miRNA, and distinguishing miRNAs from other class of ncRNAs present in the cell becomes difficult. This motivates researchers to recognize a mature miRNA from its precursor (pre-miRNA) which is longer (~80 bp) than mature miRNA and has a distinguished stem-loop secondary structure (hairpin). The hairpin secondary structure of precursor microRNA is double helical and consists of the base-pairing interactions within it. An example precursor miRNA (pre-miRNA) sequence and its corresponding hairpin loop-like secondary structure is illustrated in Figure 6.1. This hairpin structure represents base pairing interactions among nucleotides which are one of the most effective attributes for identifying miRNA from its precursors [8]. Thus the task of identifying a pre-miRNA is formulated as a binary classification

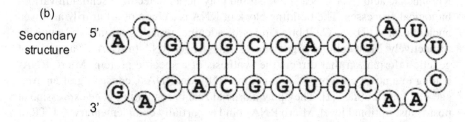

(a)

Pre-miRNA
Sequence   5' ACGUGCCACGAUUCAACGUGGCACAG 3'

(b)

Secondary
structure

FIGURE 6.1 (a) An example sequence of pre-miRNA. (b) Secondary structure of pre-miRNA sequence.

task with two classes "positive" and "negative" representing true-pre-miRNA and non-pre-miRNA (pseudo-hairpin) respectively.

6.1.2 BIOLOGY BEHIND MIRNA-TARGET (MRNA) INTERACTIONS

In molecular biology, the end-to-end (5'-end and 3'-end) chemical orientation in a single strand of nucleic acid is referred to as directionality. The 3' untranslated region (3' UTR) is the regulatory region of mRNA that is immediately followed by a translation termination codon.

A mature miRNA regulates the translation process by binding to the target site present in the 3' UTR of its corresponding mRNA. Depending on the extent of base pairing between miRNA and its target sites, it may result in the cleavage or suppression of the gene expression of target mRNA. A perfect complementary base pairing results in cleavage and an imperfect base pairing leads to translational repression. The first two to eight nucleotides starting from the 5' to the 3' UTR of a miRNA are termed a seed sequence. It is also referred to as canonical sites where the seed nucleotides perfectly complement with the target. Most of the target prediction approaches are seed centric [9], which is based on the assumption that most of the important miRNA-target interactions occur in the seed region [10]. Hence their main focus is on these canonical sites only. However, recent investigating reports on other regions (non-canonical) in a miRNA reveal the indication of many potential targets through the binding of nucleotides beyond the seed region in non-canonical sites [11, 12, 13]. Thus, it is significant to analyze the whole mature miRNA sequence to uncover the role of non-canonical sites in the target prediction process. Non-canonical sites can be 3' compensatory and centered sites with bulges or mismatches in the seed region. The candidate target site (CTS) refers to regions within a mRNA transcript with a high probability of becoming a target binding site. It is a small mRNA segment consisting of nucleotides of length k which are partially complementary to the nucleotides in the seed region at the head. Figure 6.2 represents the regulatory function of miRNA by base pairing with target sites, where blue colored lines between

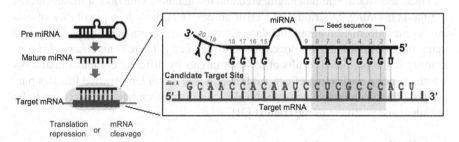

FIGURE 6.2 miRNA biogenesis and its regulatory effect by base pairing with the target sites A seed region in the miRNA consists of the first two to eight nucleotides starting from the 5' end to 3' end. The candidate target site is the small mRNA segment of length k with a high probability of becoming perfectly complemented to the seed sequence at the head [59].

nucleotides represent sequence complementing (A complements U, C complements G) formed by Watson and Crick (WC) pairing.

6.2   MIRNA SEQUENCE ANALYSIS

The study of miRNA generally stands on two defined objectives: pre-miRNA detection (i.e. identifying miRNAs from their precursors) and the prediction of miRNA targets (i.e. identifying the target sites in the mRNA that miRNA regulates). A variety of computational approaches based on machine learning algorithms, including naïve Bayes classifier (NB), neural networks (NNs), support vector machines (SVMs), and random forest (RF), have been proposed that address the objective of pre-miRNA identification and target prediction [14–18]. For both the instances of miRNA sequence analysis, these approaches require manual feature engineering to train the classifier model. In view of miRNA precursor identification, first of all, the input RNA sequence is transformed to its secondary structure using some standard methods [19, 20]. Further, numerous handcrafted features such as the occurrence of di/tri nucleotide pairs, length of loop and hairpin, the ratio of loop length to hairpin length, minimum free energy, and melting temperature [21], which encode the structural and folding stability of pre-miRNA, are extracted by domain experts. However, manual extraction of these features is cumbersome and mostly dependent on the characteristics of the data, which impacts the generalization ability and ultimately performance of the model [22, 23].

The experimental identification of miRNA targets is time consuming and costly. This promotes the reliance on various computational tools such as TargetScan [24], TargetMiner [25], TargetSpy [26], miRanda [27], MirMark [15], PITA [28], and PicTar [29] which have been suggested in literature. These tools are built upon various machine learning heuristic algorithms and empirical methods. Most of the cited tools are focused on customary features extracted from miRNA-target site interactions. Among all, seed sequence matching, sequence conservation, site accessibility, and free energy are the most commonly used features for target prediction tasks [30].

These features are human engineered and the manual extraction of these features is time-consuming, laborious, and error-prone. Thus there is a possibility of losing certain information about raw data. Consequently, it results in the generation of many false positives and a reduced score of specificity. There is an observed inconsistency in the predicted results of miRNA targets by different computational tools which creates a misleading effect. Additionally, the use of handcrafted features puts a cap on the prediction of non-canonical target sites which have been proven to be prevalent and pertinent to the context [31].

(a)   Challenges in miRNA study:

RNA sequences consist of many structural as well as biological properties, and it is very difficult to give any conclusive remark on the effectiveness of these attributes for pre-miRNA identification. The first and foremost challenge in pre-miRNA discovery is to get a relevant set of features that justify the data it represents. In this regard, researchers are either hunting for new handcrafted features or trying to combine existing features to discover new features that are meaningful for the discovery of pre-miRNAs. For the prediction of miRNA targets, a major challenge is the understanding of non-canonical miRNA-target interaction sites rather than focusing on the seed region only. Rather than relying on artificially extracted features, now the focus is on RNA sequence modeling for automatic feature learning. Another challenge addresses the class imbalance aspect in the learning process. The number of well-known (positive) miRNAs is considerably lower than the number of pseudo-hairpins (negative miRNAs). Pseudo-hairpins are normal RNA sequences with similar hairpin-like structures. However, they do not turn out into mature miRNAs. This imbalance in positive and negative samples particularly affects the supervised classifier. Concerning target prediction, the number of published true miRNA-target interactions is much higher than the number of negative interactions. Artificially generated negative target sites make the learning process difficult and mislead the classifier training.

6.3   DEEP LEARNING: CONCEPTUAL OVERVIEW

Deep learning (DL), a subfield of machine learning algorithms, is poised to reshape the learning procedure and unravel huge amounts of unstructured data [32,33]. The term "deep" in deep learning originates from the use of multiple numbers of layers in the network, which results in a deep architecture. DL models are trained by utilizing huge data sets and hierarchical levels of artificial neural networks to learn an appropriate representation of data with multiple levels of abstraction [34]. The fundamental theory behind DL is automatic feature learning from the raw data of a large training set through increasing levels of abstraction. This promotes a powerful learning process which allows the system to understand and learn complex representations from raw data through hierarchical levels of training [35]. Deep neural networks (DNNs), recurrent neural networks (RNNs), convolutional neural networks (CNNs), restricted Boltzmann machine (RBM), deep Boltzmann machine (DBM), graph convolutional networks (GCN), deep belief networks (DBN), autoencoders, and so on are the most demanding DL architectures mentioned in the literature. It's success story has made Deep Learning a pioneering technique in the diverse fields of engineering and biology [32–38]. The remarkable performance of deep learning techniques and the importance of regulatory functions of miRNA by binding to its target sites of associated mRNA motivated the exploitation of various different deep learning architectures for miRNA precursor (pre-miRNA) detection and miRNA target prediction as well. DL architectures that are most commonly used with RNA sequence data are briefly summarized below. Figure 6.3 illustrates a generalized deep learning workflow followed for RNA sequence analysis.

6.3.1   DEEP NEURAL NETWORKS (DNNS)

DNNs represent a fully connected architecture consisting of multilayer perceptron (MLP) [35], auto-encoders (AN) [39], and RBM [40]. DNN takes an input feature set

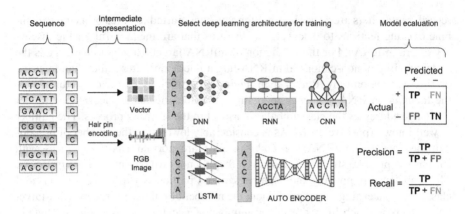

FIGURE 6.3   Comprehensive pipelines for miRNA sequence analysis.

of very high dimensions and utilizes a multi-layer architecture of neural networks to learn multiple levels of feature representation by transforming the data into a more creative and abstract component. Finally, it produces high-level relevant features for classification. Learning in the hierarchy of multiple layers facilitates the processing of high-dimensional data, exploiting and exploring the patterns in the data. Deep neural networks have also proven their performance in various bioinformatics applications including protein structure prediction, splicing pattern prediction, and so on.

6.3.2   CONVOLUTIONAL NEURAL NETWORKS (CNNs)

CNNs represent a category of deep neural network architectures which is mostly applied in the field of image and video reorganization, image classification [42], medical imaging, and speech and text processing [41]. CNN consists of multiple convolution units, a pooling unit, and finally a fully connected network. This deep architecture massively computes, integrates, and extracts an abstract feature map by establishing non-linear relationships between the input data and ground truth [43]. CNN automatically learns features from raw input features. In comparison to other deep learning models, CNN requires fewer preprocessing steps and shows impressive ability in analyzing spatial information. Thus, CNNs are most commonly applied for manipulating image data. Recently CNN has also been applied for various biological applications such as the prediction of gene expression profiling, the classification of proteins according to their functions, and protein secondary structure prediction.

6.3.3   AUTOENCODERS

An autoencoder is a type of artificial neural network which is based on the idea of unsupervised feature learning [44]. It consists of two parts: the encoder and the decoder parts. The encoder part learns an efficient encoding of data by training the network to ignore the noise. The decoder part then tries to reconstruct the original data back from the reduced encoded data with minimum reconstruction loss. The architecture of the autoencoder is mostly similar to that of MLP, with an exception that it contains the same number of neurons in the input and output layers. Variants of autoencoder models exist to improve generalization performance. Examples include denoising autoencoder (DAE) [45] and variational autoencoder (VAE) [46].

6.3.4 RECURRENT NEURAL NETWORKS (RNNs)

RNNs are mostly employed for modeling sequential or time-series data. The fundamental principle of RNN is based on a regular feed-forward network (FNN), which forms a directed graph of interconnecting nodes along a temporal sequence [47]. RNN takes a single element of time-dependent data sequence as input and processes the output using the current state and output of the previous time step. Cyclic connections among hidden units in RNN function as memory units of RNN that retain information of previously learned input and are systematically updated at each time interval. The output of the current state depends on the output of the previous states, which designate the network with a memory-like property. Recently, a variant of RNN, long short-term memory (LSTM) [48], has been successfully applied for genomics research.

6.3.5 LONG SHORT-TERM MEMORY (LSTM)

LSTM is a very special type of recurrent neural network [48], which is capable of learning order dependencies in sequence prediction tasks. The core components of each LSTM unit are a cell/node, an input gate, an output gate and a forget gate, and cyclic connection to store the state vector. Each node works for values over an interval of time, although input and output gates coordinate the information flow within the network. Recently, an attention mechanism has been introduced in LSTM which helps to learn long-term dependencies and also expedites the understanding of results [49]. LSTM gives a remarkable result in capturing long-term temporal dependencies, avoiding the optimization burdens. Hence, it is becoming a method of choice for sequence data analysis.

6.4 DEEP LEARNING: APPLICATIONS FOR PRE-MIRNA IDENTIFICATION

An increasing number of works have been proposed in the literature utilizing various deep learning architectures that have earned incredible popularity in performance with higher accuracy and faster speed than conventional methods. A brief explanation of the latest advances in pre-miRNA identification from a deep learning perspective is outlined here.

A DNN based classifier was proposed by Thomas et al. to discriminate pre-miRNAs from pseudo-hairpins [50]. A heterogeneous set of 58 features of pre-miRNAs and pseudo-hairpins including secondary structure features, sequence conservation features, and statistical features was considered for training the classifier model. The proposed architecture consisted of three hidden layers including a layer of RBM. Gibb's sampling was adopted for the adjustment of weights across layers. To handle the class imbalance problem, a k-means based sampling method was introduced. The proposed DNN model was observed with high prediction accuracy in comparison to other classical machine learning classifiers. 2D CNN was used in [51] for classifying pre-miRNAs. A down scaling layer was introduced to consolidate features and allow conversion of varying size input of CNN to a fixed-sized class prediction. RNAfold tool was utilized to predict the secondary structure [52], which was further encoded to pair matrix representation. Pairing matrix is a 2D multi-modal representation of structural, sequence, and energy features of miRNA without considering human-engineered features. Thus the shortcomings of handcrafted features were overthrown in this approach. This paired matrix is input into the convolutional network to discriminate pre-miRNA. They introduced a down scaling layer between convolution and fully connected layer facilitated the classification task with a variable size input sequence. Although pre-miRNA identification using paired matrix representation was outperforming in comparison to other cutting-edge methods, the approach demands a big data set for good generalization. Park et al. addressed the existing limitations of manual feature extraction and proposed deepMiRGene integrating information from the raw biological sequence as well as its corresponding secondary structure. RNN with LSTM unit was deployed for modeling, feature learning, and classification task of RNA sequences [53]. The generated secondary structure of input miRNA is split into a forward structure (in the direction 5' to 3' end) and a backward structure (in the direction 3' end to 5' end). Further, the miRNA sequence along with forward and backward streams were inputted to LSTM and learned in a different sequential direction. The derived secondary structure information was observed to have a higher influence on the accuracy of the model than only sequence information. Further enhancement was made to deepMiRGene in [54] and an attention mechanism was introduced in LSTM for a more enhanced and effective representation of long-term dependencies existing among the primary and secondary structure of RNA sequence. The performance of deepMiRGene was outstanding in comparison to other contemporary techniques with respect to various evaluation metrics such as sensitivity, specificity, and F-score. Wang et al. proposed a cascaded CNN-LSTM approach called CL-PMI addressing the existing challenges of manual feature engineering and class imbalance. Sequential and spatial characteristics of pre-miRNAs were considered for training the model [55]. An intermediate representation of pre-miRNA sequences and their secondary structures was obtained using one-hot encoding. The processed pre-miRNAs were inputted to CNN for the automatic learning of pre-miRNA spatial features. Further, the automatically learned and extracted spatial features of pre-miRNA were inputted to LSTM that captures the long-term dependencies of the pre-miRNAs. In order to handle the class inequality problem while training, cross-entropy loss function with focal loss was adopted. Compared to other existing methods the performance of CL-PMI was remarkable for all benchmark data sets in effectively identifying pre-miRNAs. Cordero et al. generated an intermediate graphical representation of pre-miRNA sequences that encodes sequence, structure, and thermodynamic information of pre-miRNA [56].

Then the graphical structures of pre-miRNA sequences were given to conventional CNN for training and distinguishing pre-miRNAs from pseudo-hairpins. The approach showed outstanding performance in less computational time for encoding the hairpin structure of pre-miRNAs in comparison to the existing time-consuming machine learning approaches of feature selection. Zheng et al. combined the pre-miRNA sequence information along with the predicted secondary structures of pre-miRNAs to train conventional CNN and RNN models separately for identifying pre-miRNAs [57]. These suggested models were able to learn automatically from the pattern present in the raw sequence and achieved comparable performance with other state-of-the-art existing methods.

6.5 DEEP LEARNING: APPLICATIONS FOR MIRNA TARGET PREDICTION

The capability of deep learning in autonomous feature learning and to identify patterns from raw data fascinates contemporary researchers while dealing with the challenges in discovering miRNA targets. Recently many deep learning methods have been proposed for miRNA target prediction which is proven to be effective in learning unknown features from raw RNA sequences with improved accuracy in predicting miRNA targets. Various emergent computational approaches/tools based on deep learning techniques for deciphering miRNA-target interactions are summarized below.

A CNN-based algorithm (MiRTDL) was proposed by Cheng et al. for miRNA target prediction utilizing sequence complementation, evolutionary conservation among species, and accessible features of the miRNA-target pairs [58]. A set of 20 features belonging to the above-mentioned three categories was extracted in the feature selection step and further utilized for learning of the convolution model. A constraint relaxing method was used to handle the imbalanced ratio between known positive miRNA-target interactions and negative interactions in the considered data set. It produced a balanced ratio of experimentally validated positive and negative training data sets from original positive and negative samples. The adopted relaxing method performed well for bypassing the erroneous filtering of true targets as well as handling the imbalance ratio of the positive and negative experimentally validated data sets. A CNN architecture with six layers was used to train the extracted features and predict miRNA target genes. Although the proposed CNN approach: MiRTDL achieved good accuracy in comparison to other target prediction tools, still the dependency of the approach lies in known features. Information outside these 20 selected features remains unexplored. MiRTDL employed deep learning architecture for the classification task, rather than for learning features from raw RNA sequences. As an advancement to MiRTDL, Lee et al. proposed deepTarget: a fusion of supervised and unsupervised feature learning approaches [59]. deepTarget used RNN-based autoencoders to learn the inherent representation of miRNA-target interactions without aligning the sequences. A stack of RNNs was then utilized to learn the interactions among the sequences of miRNA-target pairs.

The combined approach of sequence modeling followed by feature learning overcomes the limitation of manual feature engineering and delivers an exceptional level of accuracy compared to that of existing tools. miRAW, a DL architecture-based approach, was proposed in [60] to uncover non-canonical target sites beyond the seed region. It worked in two steps. In the first step, a novel candidate site selection method (CSSM miRAW) with more relaxed restrictions than conventional CSSM approaches was used to identify candidate target sites in the entire miRNA and 3'TR mRNA nucleotides. Proposed CSSM miRAW was able to successfully predict both canonical and a wide range of non-canonical miRNA binding sites with a higher level of accuracy. A deep ANN with eight dense hidden layers was used further for classification. Comparative investigation using independent data sets revealed that miRAW consistently outperforms with improved accuracy and a better understanding of non-canonical miRNA targets compared to other present-day prediction tools. Wen et al. proposed a deep architecture framework (DeepMirTar) for predicting human miRNA targets at site level [61]. A stacked denoising autoencoders (SDA) method was adopted for learning and accurate prediction of target mRNA genes. Seven hundred and fifty expert-designed features at three different levels including low, high, and raw levels were utilized for representing the miRNA-target site and to train the model. In comparison to other existing computational approaches utilizing various machine-learning algorithms and developed present-day tools for miRNA-target-prediction, DeepMirTar outperformed with consistent predictive accuracy. Despite this, DeepMirTar was observed with the limitation of silent dependency on four known feature types. This promotes the exploitation of both spatial as well as sequential features of raw miRNA and target sequences. A hybrid deep learning-based approach miTAR was proposed in [62] that is trained to learn both spatial as well as sequential features of miRNA and its target. CNN was designed to learn the potential features existing in the raw sequences of miRNA and its target. RNN along with LSTM unit (bidirectional LSTM) was used to learn the possible dependencies between miRNA and its target. The hybrid method of target prediction miTAR was able to predict miRNA targets with significantly higher accuracy than DeepMirTar and miRAW. Zheng et al. proposed a deep CNN-based approach to automatically learn both canonical and non-canonical interactions from primary sequences of miRNA-target site pairs [63]. The multilayer convolutional architecture consists of four layers of convolution operations followed by a max-pooling step. The learned features were then further used to predict miRNA target genes. The designed model of deep CNN architecture was able to successfully predict the miRNA target sites with great generalization ability without any manual feature extraction beforehand.

6.6   CRITICAL OBSERVATIONS AND FUTURE DIRECTIONS

Deep learning approaches automatically extract and learn features from raw data. This attribute motivates the deployment of different deep architectures for the computational prediction of human pre-miRNAs and their targets. The use of deep learning techniques defeats the time-consuming and error-prone task of manual feature engineering. Nevertheless learning from the data with an imbalanced ratio between positive and negative samples is an open challenge in the literature. But it can be tactfully overthrown using the various sampling techniques, dynamic cross entropy loss function, and constraint relaxing methods as briefed in the literature. Various deep learning models using a combination of DNN, CNN, autoencoders, RNN, and LSTM architectures have achieved outstanding performance in comparison to other contemporary methods for pre-miRNA identification and miRNA target prediction as well. From the cited literature it is observed that deep learning models require more time for training in comparison to conventional machine learning approaches. However, they achieve the best performance when various evaluation scores including F-score, true positive rate (TPR), true negative rate (TNR), Matthews correlation coefficient (MCC), and accuracy (ACC) are taken into consideration. Even so, these approaches are limited with small data set size, which is a growing challenge in omics research. Hence a promising future direction towards improved identification may be the collection of more true positive and negative data (i.e. true pre-miRNAs and pseudo miRNAs in precursor detection and positive and negative data sets for miRNA-target interactions). Additionally, the complicated interaction patterns of miRNAs and their target mRNAs are not fully understood. There are many factors concerning molecular binding that may affect miRNA-target interaction. This is still an open challenge to develop more reliable methods for miRNA target prediction. Thus exploitation of bigger neural networks such as AlexNet or ResNet towards the miRNA sequence analysis can be taken into consideration for more accurate prediction of precursor miRNAs and their target sites.

6.7 CONCLUSION

The function miRNA is prominent and potentially significant in the progression of many diseases by modulating the level of gene expression in the eukaryotic cell. Thus, identifying miRNAs and their target sites is essential for deciphering their biological functions. Computational predictions based on deep learning techniques followed by the experimental validation of predicted results certainly boost the speed of identifying true miRNAs and their target sites with a significantly high level of accuracy and reduced cost. However, the lack of reliable experimental data for building prediction models remains a limitation. This review attempts to provide a comprehensive understanding of the pipeline used for RNA sequence analysis using various deep learning architectures. Understanding the basis of the feature learning and prediction methodologies will enable the researchers for the selection and design of appropriate DL architecture for the identification of precursor miRNAs and their target genes.

BIBLIOGRAPHY

1. Bartel, D.P. "MicroRNAs: genomics, biogenesis, mechanism, and function." *Cell* 116(2) (2004): 281–297.
2. Thomas, M., J. Lieberman, and A. Lal "Desperately seeking microRNA targets." *Nature Structural & Molecular Biology* 17(10) (2010): 1169.

3. Jansson, M.D., and A.H. Lund. "MicroRNA and cancer." *Molecular Oncology* 6(6) (2012): 590–610.

4. Paul, P., A. Chakraborty, D.Sarkar, M. Langthasa, M. Rahman, M. Bari, & S. Chakraborty. Interplay between miRNAs and human diseases. *Journal of Cellular Physiology* 233(3) (2018), 2007–2018.

5. Tüfekci, K.U., M.G. Öner, R.L.J. Meuwissen, & Ş. Genç. The role of microRNAs in human diseases. In *miRNomics: MicroRNA Biology and Computational Analysis* (pp. 33–50). Humana Press, Totowa, NJ, 2014.

6. Fu, G., J. Brkić, H. Hayder, & C. Peng. MicroRNAs in human placental development and pregnancy complications. *International Journal of Molecular Sciences*, 14(3) (2013): 5519–5544.

7. Han, J., Y. Lee, K.H. Yeom, J.W. Nam, I. Heo, J.K. Rhee, ... & V.N. Kim. Molecular basis for the recognition of primary microRNAs by the Drosha-DGCR8 complex. *Cell* 125(5) (2006), 887–901.

8. Jiang, P. et al. "MiPred: classification of real and pseudo microRNA precursors using random forest prediction model with combined features." *Nucleic Acids Research* 35.suppl_2 (2007): W339–W344.

9. Schirle, N.T., J. Sheu-Gruttadauria, and I.J. MacRae. "Structural basis for microRNA targeting." *Science* 346(6209) (2014): 608–613.

10. Agarwal, V. et al. "Predicting effective microRNA target sites in mammalian mRNAs." *elife* 4 (2015): e05005.

11. Moore, M.J. et al. "miRNA–target chimeras reveal miRNA 3′-end pairing as a major determinant of Argonaute target specificity." *Nature Communications* 6(1) (2015): 1–17.

12. Broughton, J.P. et al. "Pairing beyond the seed supports microRNA targeting specificity." *Molecular Cell* 64(2) (2016): 320–333.

13. Kim, D. et al. "General rules for functional microRNA targeting." *Nature Genetics* 48(12) (2016): 1517.

14. Ding, J., X. Li, and H. Hu. "TarPmiR: a new approach for microRNA target site prediction." *Bioinformatics* 32(18) (2016): 2768–2775.

15. Menor, M. et al. "mirMark: a site-level and UTR-level classifier for miRNA target prediction." *Genome Biology* 15(10) (2014): 1–16.

16. Wang, X. "Improving microRNA target prediction by modeling with unambiguously identified microRNA-target pairs from CLIP-ligation studies." *Bioinformatics* 32.9 (2016): 1316–1322.

17. Ovando-Vázquez, C., D. Lepe-Soltero, and C. Abreu-Goodger. "Improving microRNA target prediction with gene expression profiles." *BMC Genomics* 17.1 (2016): 1–13.

18. Stegmayer, G. et al. "Predicting novel microRNA: a comprehensive comparison of machine learning approaches." *Briefings in Bioinformatics* 20.5 (2019): 1607–1620.

19. Hofacker, I.L.. Vienna RNA secondary structure server. *Nucleic Acids Research* 31(13) (2003), 3429–3431.

20. Bindewald, E., T. Kluth, & B.A. Shapiro. CyloFold: secondary structure prediction including pseudoknots. *Nucleic Acids Research* 38(suppl_2) (2010): W368–W372.

21. Xue, C., F. Li, T. He, G.P. Liu, Y. Li, & X. Zhang. Classification of real and pseudo microRNA precursors using local structure-sequence features and support vector machine. *BMC Bioinformatics* 6(1) (2005): 310.

22. Demirci, M.D.S., J. Baumbach, & J. Allmer. On the performance of pre-microRNA detection algorithms. *Nature Communications* 8(1) (2017): 1–9.

23. de ON Lopes, I., A. Schliep, & A.C.D.L. de Carvalho. The discriminant power of RNA features for pre-miRNA recognition. *BMC Bioinformatics* 15(1) (2014), 124.

24. Lewis, Benjamin P., et al. "Prediction of mammalian microRNA targets." *Cell* 115(7) (2003): 787–798.

25. Bandyopadhyay, S., and R. Mitra. "TargetMiner: microRNA target prediction with systematic identification of tissue-specific negative examples." *Bioinformatics* 25(20) (2009): 2625–2631.

26. Sturm, M. et al. "TargetSpy: a supervised machine learning approach for microRNA target prediction." *BMC Bioinformatics* 11(1) (2010): 292.

27. John, B. et al. "Human microRNA targets." *PLoS Biol* 2(11) (2004): e363.

28. Kertesz, M. et al. "The role of site accessibility in microRNA target recognition." *Nature Genetics* 39(10) (2007): 1278–1284.

29. Krek, A. et al. "Combinatorial microRNA target predictions." *Nature Genetics* 37(5) (2005): 495–500.

30. Peterson, S.M. et al. "Common features of microRNA target prediction tools." *Frontiers in Genetics* 5 (2014): 23.

31. Agarwal, V. et al. "Predicting effective microRNA target sites in mammalian mRNAs." *elife* 4 (2015): e05005.

32. LeCun, Y., Y. Bengio, and G. Hinton. "Deep learning." *Nature* 521(7553) (2015): 436–444.

33. Koumakis, L. "Deep learning models in genomics; are we there yet?." *Computational and Structural Biotechnology Journal* (2020).

34. Bengio, Y. *Learning Deep Architectures for AI*. Now Publishers Inc, 2009.

35. Goodfellow, I., Y. Bengio, and A. Courville. *Deep Learning*. MIT press, 2016.

36. Golkov, V. et al. "3d deep learning for biological function prediction from physical fields." arXiv preprint arXiv:1704.04039 (2017).

37. Spencer, M., J. Eickholt, and J. Cheng. "A deep learning network approach to ab initio protein secondary structure prediction." *IEEE/ACM Transactions on Computational Biology and Bioinformatics* 12(1) (2014): 103–112.

38. Golkov, Vladimir et al. "Protein contact prediction from amino acid co-evolution using convolutional networks for graph-valued images." *Advances in Neural Information Processing Systems* (2016).

39. Hinton, G.E., and R.R. Salakhutdinov. "Reducing the dimensionality of data with neural networks." *Science* 313(5786) (2006): 504–507.

40. Hinton, G.E., and T.J. Sejnowski. "Learning and relearning in Boltzmann machines." *Parallel Distributed Processing: Explorations in the Microstructure of Cognition* 1 (1986): 282–317, 2.

41. Krizhevsky, A., I. Sutskever, and G.E. Hinton. "Imagenet classification with deep convolutional neural networks." *Advances in Neural Information Processing Systems* (2012).

42. Li, H. et al. "A convolutional neural network cascade for face detection." *Proceedings of the IEEE Conference on Computer Vision and Pattern Recognition* (2015).

43. Gu, J., et al. "Recent advances in convolutional neural networks." *Pattern Recognition* 77 (2018): 354–377.

44. Baldi, P.. "Autoencoders, unsupervised learning, and deep architectures." *Proceedings of ICML Workshop on Unsupervised and Transfer Learning* (2012).

45. Vincent, P., et al. "Extracting and composing robust features with denoising autoencoders." *Proceedings of the 25th International Conference on Machine Learning* (2008).

46. Kingma, D.P., and M. Welling. "Auto-encoding variational bayes." arXiv preprint arXiv:1312.6114 (2013).

47. Williams, R.J., and D. Zipser. "A learning algorithm for continually running fully recurrent neural networks." *Neural Computation* 1(2) (1989): 270–280.

48. Hochreiter, S., and J. Schmidhuber. "Long short-term memory." *Neural Computation* 9(8) (1997): 1735–1780.
49. Xu, K. et al., Show, Attend and Tell: Neural Image Caption Generation with Visual Attention. In *ICML*, volume 14, pages 77–81, 2015.
50. Thomas, J., S. Thomas, and L. Sael. "DP-miRNA: An improved prediction of precursor microRNA using deep learning model." *2017 IEEE International Conference on Big Data and Smart Computing (BigComp)*. IEEE, 2017.
51. Do, B.T. et al. "Precursor microRNA identification using deep convolutional neural networks." *BioRxiv* (2018): 414656.
52. Hofacker, I.L. "Vienna RNA secondary structure server." *Nucleic Acids Research* 31(13) (2003): 3429–3431.
53. Park, S. et al. "deepMiRGene: Deep neural network based precursor microrna prediction." arXiv preprint arXiv:1605.00017 (2016).
54. Park, S. et al. "Deep recurrent neural network-based identification of precursor micrornas." *Advances in Neural Information Processing Systems*. 2017.
55. Wang, H., et al. "CL-PMI: A precursor microRNA identification method based on convolutional and long short-term memory networks." *Frontiers in Genetics* 10 (2019): 967.
56. Cruz, J.A.C., V. Menkovski, and J. Allmer. "Detection of pre-microRNA with Convolutional Neural Networks." *bioRxiv* (2020): 840579.
57. Zheng, X. et al. "Deep neural networks for human microRNA precursor detection." *BMC Bioinformatics* 21(1) (2020): 1–7.
58. Cheng, S. et al. "MiRTDL: a deep learning approach for miRNA target prediction." *IEEE/ACM Transactions on Computational Biology and Bioinformatics* 13(6) (2015): 1161–1169.
59. Lee, B. et al. "deepTarget: end-to-end learning framework for microRNA target prediction using deep recurrent neural networks." *Proceedings of the 7th ACM International Conference on Bioinformatics, Computational Biology, and Health Informatics*. 2016.
60. Pla, A., X. Zhong, and S. Rayner. "miRAW: A deep learning-based approach to predict microRNA targets by analyzing whole microRNA transcripts." *PLoS Computational Biology* 14.7 (2018): e1006185.
61. Wen, M., et al. "DeepMirTar: a deep-learning approach for predicting human miRNA targets." *Bioinformatics* 34(22) (2018): 3781–3787.
62. Gu, T., et al. "miTAR: a hybrid deep learning-based approach for predicting miRNA targets." *bioRxiv* (2020).
63. Zheng, X., et al. "Prediction of miRNA targets by learning from interaction sequences." *Plos one* 15(5) (2020): e0232578.

7 Role of Machine Intelligence in Cancer Drug Discovery and Development

*Kaushik Kumar Bharadwaj, Bijuli Rabha,
Gaber El-Saber Batiha, Debabrat
Baishya, and Arabinda Ghosh*

CONTENTS

7.1   INTRODUCTION

Cancer has become the leading health trouble worldwide because of its gradually increasing rate of frequency and mortality. The GLOBOCAN 2018 database (http://gco.iarc.fr/l), also known as Global Cancer Observatory (GCO) a web-based platform presenting global cancer statistics, estimates 18.1 million cases and 9.6 million mortalities in the year 2018 throughout the world with lung, breast, and colorectum the highest ranked. The discovery and development of life-saving drugs are the most significant and translational science exercises that contribute to the healthiness of mankind. The drug development process is a very long and delayed process usually described as finding a needle in a haystack, requiring many steps, beginning with drug target identification, potential drug identification, in vitro drug testing, toxicity studies, in vivo testing in human cell lines and animal models, and lastly clinical trials in humans [1]. The economic investment in drug design and development processes is so high and costs billions with a relatively high risk of failure in clinical trials, and it also takes almost 15 years for a novel drug molecule to come to the market [2]. Drug development by pharmaceutical industries faces many key

DOI: 10.1201/9781003126164-7

challenges related to finding a correct drug target, the screening of potential drugs from a huge database of molecules, toxicity, side effects of drugs, and appropriate drug doses for a specific patient. The current improvement in high-performance computational power and the development of computer-aided drug designing techniques (CADD) has dramatically given thrust to the drug discovery and development pipeline to hasten the progression as well as also reducing the cost [3]. The current advances of computational intelligence like artificial intelligence (AI) and machine learning (ML) offer a great opportunity for researchers for drug design and development processes in the pharmaceutical industry. The availability and application of graphical processing units (GPU) hardware have quickened the computational speed for efficient drug discovery. Due to the improvement in computer hardware and the availability of a very large number of datasets (e.g., genomics, proteomics, transcriptomics, metabolomics, protein-protein networks, etc.), AI- and ML-guided drug designing and development can now be envisioned. Databases like PubChem (https://pubchem.ncbi.nlm.nih.gov/), ChEMBL (https://www.ebi.ac.uk/chembl/), and ZINC (https://zinc.docking.org/) have millions of molecules that are extremely valuable resources for AI and ML in drug designing and development applications for cancer therapy [4]. ML, a subfield of AI, performs virtual screening of potential drug molecules from databases by using a statistical method called QSAR modeling [5]. ML algorithms can also perform secondary assays by predicting the pharmacokinetic properties like absorption, distribution, metabolism, excretion, and toxicity (ADMET) for better screening of drugs with potentially fewer side effects. AI has become a potential tool used all over in the various processes of drug designing and development strategies, such as the study of drug repurposing [6], the prediction of practicable routes for the synthesis of potential drug compounds, the prediction of toxicity and pharmacological characteristics, the prediction of drug interaction with the target, the identification of novel therapeutic targets, and the forecasting of cancer cells' sensitivity to drugs [7]. The accurate prediction by the AI algorithmic tool would successfully improve the clinical trial results in cancer research [8].

7.2 ARTIFICIAL INTELLIGENCE (AI), MACHINE LEARNING (ML), AND DEEP LEARNING (DL) IN DRUG DISCOVERY AND DEVELOPMENT

AI, also sometimes called machine intelligence, is the branch of engineering that creates machines that mimic human intelligence. John McCarthy in 1956 at the Dartmouth Conference coined the term "artificial intelligence." AI algorithms mimic the human cognitive functions which are associated with learning and problem-solving behavior [9]. ML is a subset of AI which learns and improves without programming by a human being [10]. It uses algorithms which use previous experimental data (training set) to discover the functional patterns of the drug molecules. Various ML predicting tools are becoming popular in drug design and development research for pharmacological and drug repurposing predictions [11]. DL is a subset of ML also known as the new or next-generation machine learning that allows multiple processing layers from huge datasets, and DL has shown tremendous potential in successfully identifying potent drug molecules and also in predicting their pharmacological properties [12]. In Figure 7.1 the schematic representation and the association between AI, ML, and DL are depicted.

In 2014, the Dialogue on Reverse Engineering Assessment and Methods (DREAM) project in combination with the National Cancer Institute (NCI) (https://www.cancer .gov/) developed a computational model to successfully predict drug responsiveness in breast cancer using genomic and proteomic data [13]. An anticancer drug sensitivity prediction model using elastic net regression, an ML method, was developed by taking genomic data of 1005 cancer cell lines and drug-responsive data of 265 anticancer compounds [14]. ML algorithms were able to efficiently predict the drug sensitivity before treatment within various cancer patients such as ovarian, gastric, and endometrial cancers [15, 16, 17]. AI is perfectly able to optimize the synergistic process of anticancer drugs. An AI named "CURATE.AI" developed by the National University of Singapore, based on deep learning, successfully determined the dose of anticancer drugs enzalutamide and ZEN-3694 [18]. AI accelerates the prediction process of the sensitivity of anticancer drugs, synergistic responses, and the tolerance of drug doses in the management of cancer chemotherapy [19]. Deep learning-based algorithmic methods named "DeepDTIs" and plain deep classifier and conventional machine learning-based "DeepCCI" can accurately predict the potential interactions of the drug with the proteins and chemical–chemical interactions which assist the exploration of the pharmacokinetic properties of drugs [20, 21]. Random forest (RF), an ensemble modeling method of ML, was used in predicting toxicity [22], predicting protein-ligand binding interaction [23], and identifying novel drug targets [24]. In Figure 7.2 the various techniques used in machine learning for drug discovery are

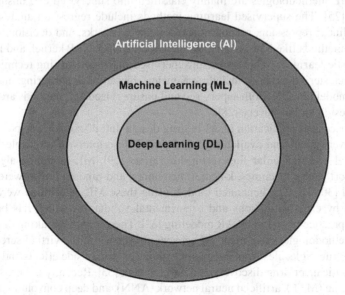

FIGURE 7.1 Schematic representation showing the association between AI, ML, and DL.

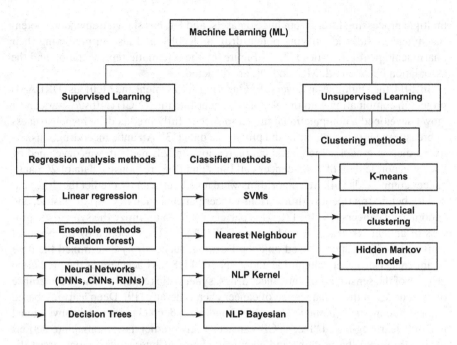

FIGURE 7.2 Machine learning (ML) techniques used in drug discovery and development process (support vector machine (SVM), deep neural network (DNN), convolutional neural network (CNN), recurrent neural network (RNN), and natural language processing (NLP)).

shown. ML methodologies are mainly classified into supervised and unsupervised learning [25]. The supervised learning methods include regression analysis methods like linear regression, random forests, neural networks, and decision trees and classifier methods like SVMs, k-nearest neighbor (kNN), NLP kernel, and Bayesian probabilistic learning. Whereas the unsupervised machine learning techniques are based on clustering methods including K-means, hierarchical clustering, and hidden Markov model [26, 27]. Both supervised and unsupervised ML methods are utilized in drug design and discovery [28].

The successful application of AI in drug design and discovery was possible due to the development and availability of chemical descriptors of molecules such as topological and molecular fingerprint descriptors [29]. ML modeling algorithms, viz. support vector machines, k-nearest neighbors, and random forest, were usually employed [30, 31]. The generated models using these ML algorithms were always validated by cross-validations and experimental validations which has become a standard practice model in QSAR modeling [32]. These groundbreaking discoveries of ML methodologies have made AI achieve milestones in the virtual screening of potential drug molecules from databases with fewer human side effects and less toxicity for anticancer drug discovery and development [33]. Recently the use of multi-task learning (MTL), artificial neural network (ANN), and deep convolutional neural network (DCNN) protein-ligand scoring functions to predict the binding poses, binding affinity, and bioactivity of small molecules was developed [34]. An online virtual screening web tool named "MLViS" was developed by utilizing ANN for screening drug-like molecules during the early phase [35]. Computational chemistry research areas like molecular dynamics simulations (MD) or quantum mechanics (QM)/molecular mechanics (MM) are important methods in predicting the binding interactions of ligand (drug) with target proteins and to measure the atomic vibration of the biomolecular system [36]. The combinatorial powers of AI with computational chemistry highly impact the efficiency in simulation capacity for the better [37, 38]. Various pharmaceutical industries are currently focused on AI-based drug design and development in their research and development plan of action in collaboration with academic institutions and startups [39]. Recursion Pharmaceuticals, in collaboration with Takeda Pharmaceutical Ltd, have identified novel preclinical compounds for a rare disease within 1.5 years which is much quicker than the conventional drug discovery pipeline. Moreover, Merck has developed a DL-based algorithm which can predict native protein folding within a day [40]. In Table 7.1 AI-based tools used in drug discovery are mentioned. AlphaFold can create high-accuracy protein structure prediction by neural network, which might help in detailed analyzing of the functions of structurally unsolved proteins [41]. ML approaches like random forests

TABLE 7.1
AI Tools Used in Drug Discovery

Sl. No.	Tools	Uses	Websites	References
i.	AlphaFold	3D structure prediction of proteins	https://deepmind.com/blog/alphafold	[41]
ii.	DeepChem	Drug discovery, quantum chemistry	https://github.com/deepchem/deepchem	[43]
iii.	DeepTox	Toxicity predictions	www.bioinf.jku.at/research/DeepTox	[44]
iv.	DeepNeuralNet-QSAR	Prediction of activities	https://github.com/Merck/DeepNeuralNet-QSAR	[45]
v.	DeltaVina	Study of protein–ligand interactions	https://github.com/chengwang88/deltavina	[46]
vi.	NNScore	Study of protein–ligand interactions	http://rocce-vm0.ucsd.edu/data/sw/hosted/nnscore/	[47]
vii.	PPB2	Polypharmacology prediction	http://ppb2.gdb.tools/	[48]
viii.	SCScore	Evaluation of the complexity of a molecule during synthesis	https://github.com/connorcoley/scscore	[49]
ix.	SIEVE-Score	Structure-based virtual screening	https://github.com/sekijima-lab/SIEVE-Score	[50]
x.	Potential Net	Ligand binding prediction	https://pubs.acs.org/doi/full/10.1021/acscentsci.8b00507	[51]

and elastic net regression methods are developed to predict various drug effects on specific cancer cells. To investigate the specific molecular features to predict therapeutic drug responses, various anticancer drug responses on numerous cancer cell lines were studied and analyzed to see how gene expression, DNA methylation, copy number variations, and somatic mutation of cancer cells affect different drug responses [42]. This type of methodology cut the huge cost of screening anticancer drugs by *in vitro* methods.

7.3 CHALLENGES TO OVERCOME

AI and machine learning algorithms are incessantly propagating and are often described as game-changing technologies. They have slowly but steadily started to transform the pharmaceutical industry throughout the last five years. In the modern era of digital health care and AI, the pharmaceutical industry and the pharmaceutical manufacturing giants are desperately in need of new transformative technology. DL methods have been enormously popular over the last decade and have been widely used to establish an AI in nearly every territory, particularly because it has been achieving its ostentatious record on the computational field. AI possesses high logical reasoning proficiencies with autonomous cognitive potential that can mimic the human brain's cogitation efficiently. AI technologies, viz. machine learning, can have a colossal impact and can greatly boost the current manner of anticancer drug discovery research [52]. Nonetheless, at the present moment, AI also has its own limitations. Given the obvious use of Big Data and AI in drug research, there exists a domain that still cannot thrive significantly. The sum of publicly accessible chemical and biological data has risen markedly in the last decade. However, ADMET and therapeutic approaches to drug treatment are the obvious domains in which there is currently inadequate research evidence and public data accessible to achieve the best out of sophisticated AI algorithms [53]. Perhaps, oncology is the clinical field that has gained the most from Big Data, not least because of the volume of data and information currently accessible to the general public [54]. The quality of the data is also a general issue. Efforts to properly treat, standardize, and clean up the available data will be difficult, especially before improved ML algorithms are established that are properly able to tackle existing noise and other data limitations.

AI provides a promising conduit to the next phase in cancer treatment, reconciling enormous volumes and diverse data forms into actionable intervention. The siloed usages of small segments of the cancer therapy system are among the hurdles to implementation. For example, AI-optimized drugs that are sub-optimally paired with, or improperly dosed with, other treatments are unlikely to significantly boost patient results [55]. To address this challenge in oncology, the comprehensive adoption of AI across the continuum of research, growth, and administration would be needed. Its impending downstream utilization includes augmenting the determination of personalized therapy by a customizing process that incorporates numerous therapeutic stratagems. For example, AI-optimized radiation treatment dosing can theoretically be paired with AI-driven drug administration, which retains comprehensive tumor size regulation. Therefore, the particular difficulties in the implementation of ML in health care delivery are that it needs pre-processing data, model testing, and model refinement concerning the actual clinical issue [56]. Also pivotal are the ethical implications, which incorporate medico-legal implications, doctors' understanding of ML tools, and information confidentiality, protection, and safety. Ultimately, the systematic introduction of AI into clinical oncology practice could increase treatment quality and reduce the cost of health care. As AI continues to be validated and a road towards mainstream use is established, the capacity for redefining cancer therapy clinical practices is becoming apparent.

7.4 FUTURE PROSPECTS AND CONCLUSION

The increased development of Big Data has created many new biotechnology firms to leverage AI approaches in drug research, such as Benevolent AI, Atomwise, and Exscientia. There is a strong incentive in the business arena to harness the greater abundance of data and algorithms. So, it can be rightly assumed that the Big Data revolution is here to stay and rule the next era. The rationale behind the use of Big Data and AI is to promote improved drug discovery decision-making. So far, it is clear that Big Data's greatest effect and gain are seen in fields where it forms an indispensable constituent of the spectrum of drug development, for instance as observed currently in target selection and drug design [55]. In addition to making Big Data and AI techniques accessible, proper training and preparation are critical for the upcoming generation of drug discoverers to take proper and adequate advantage of it. This should ensure that the effective methods and tools are optimally available for the drug discoverers to utilize. Public Big Data and its optimum alignment with the internal data of an organization are projected to become a progressively important tool in drug research, along with the improvement of refined algorithms for distilling insight and information from complicated multidisciplinary data. These methods must be embraced as an important part of the whole drug discovery process and proceed into pharmaceutical research and treatment to optimize the advantages of Big Data and AI [57]. In developing drug combinations, AI can play a crucial role without relying on synergy-based models or predicted synergies between various drug targets and pathways. This will greatly expand the number of medications available for treatment and detect unusual formulations that meet clinical expectations. The possible drugs and dosage ranges are too expensive for thorough evaluation when considering the dosage of each candidate drug, too. AI can therefore easily overcome broad gaps in medications and dosage parameters. The quadratic phenotypic optimization platform (QPOP), for example, uses quadratic parabolic relationships to visually equate a set of inputs (e.g., medications and doses) with ideal outputs (e.g., preclinical tumor decreasing with reduced toxicity). This association significantly diminishes the number of experiments and data required to recognize the drugs and doses which optimize the combination. Besides, QPOP is also agnostic to disease pathways and mechanisms, drug targets, and interaction with drugs. The applications to AI have the potential to influence numerous aspects of cancer therapy. These include drug discovery and development among others, and these drugs are scientifically tested and eventually delivered at the point of repair. These procedures are already arduous and time-consuming. Therapies can frequently contribute to varying effects of care between patients. The integration of AI and cancer treatment has led to many answers to these problems. AI technologies extending from ML to ANN will expedite drug development, leverage biomarkers to fit patients reliably to clinical trials, and fully tailor cancer treatment using only the patient's own data. These findings hint that AI-enhanced, functional cancer treatment could be on the way to take over traditional drug discovery techniques.

ABBREVIATIONS

ADMET: absorption, distribution, metabolism, excretion, and toxicity
AI: artificial intelligence
CADD: computer-aided drug designing techniques
CNN: convolutional neural network
DNN: deep neural networking
DREAM: Dialogue on Reverse Engineering Assessment and Methods
GPU: graphical processing units
IARC: International Agency for Research on Cancer
ML: machine learning
MM: molecular mechanics
NCI: National Cancer Institute
NLP: natural language processing
QM: quantum mechanics
QPOP: quadratic phenotypic optimization platform
RNN: recurrent neural network
SVM: support vector machine

BIBLIOGRAPHY

1. D. Vohora, and G. Singh, *"Pharmaceutical Medicine and Translational Clinical Research,"* (Academic Press), Nov. 2017. https://www.elsevier.com/books/pharmaceutical-medicine-and translational-clinical-research/vohora/978-0-12-802103-3.
2. H.M. Kantarjian et al., "Cancer Research in the United States: A Critical Review of Current Status and Proposal for Alternative Models," *Cancer*, vol. 124, pp. 2881–2889, Jul. 2018.
3. F. Zhong et al., "Artificial intelligence in drug design," *Sci China Life Sci.*, Vol. 61, pp. 1191–1204, Oct. 2018.
4. S. Ekins et al., "Exploiting machine learning for end-to-end drug discovery and development," *Nature Materials*, vol. 18, pp. 435–441, Apr. 2019.
5. F. Ghasemi et al., "Neural network and deep-learning algorithms used in QSAR studies: merits and drawbacks," *Drug Discov. Today*, vol. 23, pp. 1784–1790, Jun. 2018.
6. K.K. Mak, and M.R. Pichika,"Artificial intelligence in drug development: present status and future prospects," *Drug Discovery Today*, vol. 24, pp. 773–780, Mar. 2019.
7. M.P. Menden et al., "Machine learning prediction of cancer cell sensitivity to drugs based on genomic and chemical properties," *PLoS One*, vol. 8, no. e61318, Apr. 2013.
8. J.L. Perez-Gracia et al., "Strategies to design clinical studies to identify predictive biomarkers in cancer research," *Cancer Treat. Rev.*, vol. 53, pp. 79–97, Feb. 2017.

9. S. Russell, and P. Norvig, *"Artificial Intelligence: A Modern Approach* (4th edn),"Pearson. 2019.

10. A. Plante et al., "A machine learning approach for the discovery of ligand-specific functional mechanisms of GPCRs,"*Molecules*, vol. 24, no. 2097, Jun. 2019.

11. A. Aliper *et al.*, "Deep learning applications for predicting pharmacological properties of drugs and drug repurposing using transcriptomic data," *Mol Pharm.*, vol. 13, pp. 2524–2530, May 2016.

12. J.P. Hughes et al., "Principles of early drug discovery," *Br. J. Pharmacol.*, vol. 162, pp. 1239–1249, Mar. 2011.

13. J.C. Costello et al., "A community effort to assess and improve drug sensitivity prediction algorithms," *Nat. Biotechnol.*, vol. 32, no. 12, Dec. 2014.

14. Y. Wang et al., "Systematic identification of non-coding pharmacogenomic landscape in cancer," *Nat. Commun.*, vol. 9, no. 3192, Aug. 2018.

15. M.A. Hossain *et al.*, "Machine learning and bioinformatics models to identify gene expression patterns of ovarian cancer associated with disease progression and mortality," *J. Biomed. Inform.*, vol. 100, no. 103313, Dec. 2019.

16. J. Taninaga et al., "Prediction of future gastric cancer risk using a machine learning algorithm and comprehensive medical check-up data: a case-control study," *Sci. Rep.*, vol. 9, no. 12384, Aug. 2019.

17. E. Gunakan et al., "A novel prediction method for lymph node involvement in endometrial cancer: machine learning," *Int. J. Gynecol. Cancer*, vol. 29, pp. 320–324, Feb. 2019.

18. A.J. Pantuck et al., "Modulating BET bromodomain inhibitor ZEN-3694 and enzalutamide combination dosing in a metastatic prostate Cancer patient using CURATE.AI, an artificial intelligence platform," *Adv. Ther.*, vol. 1, no. 1800104, Aug. 2018.

19. G. Chen et al., "Predict effective drug combination by deep belief network and ontology fingerprints," *J. Biomed. Inform.*, vol. 85, pp. 149–154, Sep. 2018.

20. M. Wen et al., "Deep-learning-based drug-target interaction prediction," *J. Proteome. Res.*, vol. 16: 1401–1409, Mar. 2017.

21. S. Kwon, and S. Yoon, "DeepCCI: end-to-end deep learning for chemical-chemical interaction prediction," arXiv 1704.08432, 2017.

22. P. Mistry et al., "Using random forest and decision tree models for a new vehicle prediction approach in computational toxicology," *Soft. Comput.*, vol. 20, pp. 2967–2979, Nov. 2016.

23. Y. Wang et al., "A comparative study of family-specific protein-ligand complex affinity prediction based on random forest approach," *J. Comput. Aid. Mol. Des.*, vol. 29, pp. 349–360, Apr. 2015.

24. P. Kumari et al., "Identification of human drug targets using machine-learning algorithms," *Comput. Biol. Med.*, vol. 56, pp. 175–181, Nov. 2014.

25. Y.C. Lo et al., "Machine learning in chemoinformatics and drug discovery," *Drug Discovery Today.*, vol. 23, pp. 1538–1546, Aug. 2018.

26. V. Svetnik et al., "Random forest: a classification and regression tool for compound classification and QSAR modeling," *J. Chem. Inf. Comput. Sci.*, vol. 43, pp. 1947–1958, Nov. 2003.

27. I.I. Baskin et al., "A renaissance of neural networks in drug discovery," *Expert Opin. Drug Discov.*, vol. 11, pp. 785–795, Jul. 2016.

28. M.W. Libbrecht, W.S. Noble, "Machine learning applications in genetics and genomics," *Nat Rev Genet.*, vol. 16, pp. 321–332, May, 2015.

29. R. Gozalbes, J.P. Doucet, and F. Derouin, "Application of topological descriptors in QSAR and drug design: history and new trends," *Curr. Drug Targets Infect. Disord.*, vol. 2, pp. 93–102, Mar. 2002.

30. B. Sprague et al., "Design, synthesis and experimental validation of novel potential chemopreventive agents using random forest and support vector machine binary classifiers," *J. Comput.-Aided Mol. Des.*, vol. 28, pp. 631–646, May 2014.

31. L. Breiman, "Random forests," *Mach. Learn.*, vol. 45, pp. 5–32, Oct. 2001.

32. A. Tropsha, "Best practices for QSAR model development, validation, and exploitation," *Mol. Informat.*, vol. 29, pp. 476–88, Jul. 2010.

33. S. Ekins et al., "Towards a new age of virtual ADME/TOX and multidimensional drug discovery," *Mol. Divers.*, vol. 5, pp. 255–275, Dec. 2000.

34. H.M. Ashtawy, and N.R. Mahapatra, "Task-specific scoring functions for predicting ligand binding poses and affinity and for screening enrichment," *J. Chem. Inf. Model.*, vol. 58, pp. 119–133, Jan. 2018.

35. S. Korkmaz, G. Zararsiz, and D. Goksuluk, "MLViS: a web tool for machine learning-based virtual screening in early-phase of drug discovery and development," *PLoS ONE*, no. 10: e0124600, Apr. 2015.

36. M. Wang et al., "Predicting relative binding affinity using non equilibrium QM/MM simulations," *J. Chem. Theory Comput.*, vol. 14, pp. 6613–6622, Dec. 2018.

37. L. Li et al., "Understanding machine-learned density functionals," *International Journal of Quantum Chemistry*, vol. 116, pp. 819–833, Jun. 2016.

38. M. Rupp, "Machine learning for quantum mechanics in a nutshell," *International Journal of Quantum Chemistry*, vol. 115, pp. 1058–1073, Aug. 2015.

39. K.K. Mak, and M.R. Pichika, "Artificial intelligence in drug development: present status and future prospects," *Drug Discov. Today*, vol. 24, pp. 773–780, Mar. 2019.

40. A. Bada, "World's oldest pharmaceutical Merck wins new AI & block chain patent,"*BTCNN.*, 1 February, 2019.

41. Robert F. Service, "Google's DeepMind aces protein folding," Dec. 6, 2018. www.sciencemag.org/news/2018/12/google-s-deepmind-aces-protein-folding.

42. Iorio, F. et al. A Landscape of Pharmacogenomic interactions in Cancer. *Cell* 166, 740–754 2016.

43. B. Ramsundar et al., *"DeepLearning for the Life Sciences,"* O'Reilly Media, Apr. 2019.

44. A. Mayr et al., "DeepTox: toxicity prediction using deep learning, " *Front. Environ. Sci.* vol. 3, no. 80, Feb. 2016.

45. Y. Xu et al., "Demystifying multitask deep neural networks for quantitative structure–activity relationships," *J. Chem. Inf.Model.*, vol. 57, pp. 2490–2504, Sep. 2017.

46. C. Wang, and Y. Zhang, "Improving scoring-docking-screening powers of protein–ligand scoring functions using random forest," *J. Comput. Chem.*, vol. 38, pp. 169–177, Jan. 2017.

47. J.D. Durrant, and J.A. Mc Cammon, "NNScore2.0: a neural-network receptor–ligand scoring function," *J. Chem. Inf. Model.*, vol. 51, pp. 2897–2903, Oct. 2011.

48. M. Awale, and J.L. Reymond, "Polypharmacology browserPPB2: target prediction combining nearest neighbors with machine learning," *J. Chem. Inf. Model.*, vol. 59, pp. 10–17, Jan. 2019.

49. C.W. Coley et al., "S C Score: synthetic complexity learned from a reaction corpus," *J. Chem. Inf.Model.*, vol. 58, pp. 252–261, Jan. 2018.

50. N. Yasuo, and M. Sekijima, "Improved method of structure-based virtual screening via interaction-energy-based learning," *J. Chem. Inf. Model.*, vol. 59, pp. 1050–1061, Feb. 2019.

51. E.N. Feinberg et al., "Potential Net for molecular property prediction," *ACS Cent. Sci.*, vol. 4, pp. 1520–1530, Nov. 2018.

52. G. Liang et al., "The emerging roles of artificial intelligence in cancer drug development and precision therapy," *Biomedicine & Pharmacotherapy*, vol. 128, no. 110255, Aug. 2020.

53. P. Workman, A.A. Antolin, and B. Al-Lazikani, "Transforming cancer drug discovery with Big Data and AI," *Expert Opinion on Drug Discovery*, vol. 14, pp. 1089–1095, Jul. 2019.

54. K.Y. Ngiam, and W. Khor, "Big data and machine learning algorithms for health-care delivery," *The Lancet Oncology*, vol. 20, pp. e262–e273, May 2019.

55. H. Dean, "Artificial intelligence in cancer therapy," *Science*, vol. 367, pp. 982–983, Feb. 2020.

56. W. Duch, K. Swaminathan, and J. Meller, "Artificial intelligence approaches for rational drug design and discovery," *Curr. Pharm. Des.*, vol. 13, pp. 1497–508, 2007.

57. Y. Jing et al., "Deep learning for drug design: an artificial intelligence paradigm for drug discovery in the big data era," *The AAPS Journal*, vol. 20, no. 58, Mar. 2018.

8 Genome and Gene Editing by Artificial Intelligence Programs

Imran Zafar, Aamna Rafique,
Javaria Fazal, Misbah Manzoor,
Qurat Ul Ain, and Rehab A. Rayan

CONTENTS

DOI: 10.1201/9781003126164-8

8.1   INTRODUCTION

Genome editing is a tool that enables experts to alter the DNA of many organisms, such as plants, bacteria, and animals [1]. DNA editing can lead to alterations in biochemical properties, such as facial structure and chronic diseases. Researchers use different approaches to do anything like this; such techniques work like scissors, cutting DNA at a particular place [2]. It is then necessary for investigators to erase, add, or replace DNA where it has been removed. Developments in the production of genomic instruments, such as zinc finger proteins, TALE, and more recently, the CRISPR-Cas9 process, allow all to study the genomes of different organisms at a conceptual and functional level with unparalleled sensitivity and flexibility [3].

Genome editing is a mixture of processes that provide an opportunity for researchers to modify an organism's DNA. The above strategies allow genetic material modification at multiple genome locations, either by addition or deletion. There were numerous editing modes added. CRISPR/Cas-9 has a lot of short palindrome iterations, and one of them is CRISPR-associated protein 9. The CRISPR-Cas9 system is more efficient, precise, and fast, and cheaper than any other current editing methods, making it the leading cause of scientific community controversy [6].

Genome editing is a critical player in the diagnosis of and intervention in biological-associated diseases [7]. Much effort is currently underway on genome editing to clarify disorders utilizing various cellular and animal templates. Researchers continue to believe that this technique is useful for clinical operations. Many medical conditions are now being studied, including rare condition genetic diseases, hypothyroidism, and sickle cell disease. It also means that more severe disorders, such as cancer, cardiac disease, mental illness, or human immunodeficiency virus (HIV), are adequately managed and monitored. Normative issues arise when genome editing is done using techniques such as CRISPR-Cas9 [3] to alter biological datasets.. Genome editing can be performed in germline cells (sperm, eggs or embryos) to induce heritable genetic changes or in somatic cells (other cells) to induce. Such modifications only affect these tissues, which cannot be produced. In egg or sperm cells (germ cells) or embryo genes, changes can also be transferred to the next generation. Modifying the sperm and egg genotypes;embryonic disease raises amoral issue , (for example, height or intelligence). In some areas of the world, gametes and oocytes' production is strictly unlawful based on theory and security measures.

FIGURE 8.1 Genomic data and AI program implementation for output optimization.

Drugs can be changed all over the world through artificial intelligence and genome editing. While there is no new technology, CRISPR's invention will hopefully produce all therapeutic effects through editing and advances in deep AI learning. In medical practice, using these techniques will contribute to their potential collective use, posing new and troubling ethical concerns, both legal and social, as seen in Figure 8.1.

The combination of AI and CRISPR will "improve" human health if it is possible to address ongoing technological challenges [8].. Living cells are used to construct biological computers. Chemical inputs and other biologically generated substances such as proteins and DNA are used instead of electrical equipment and signalling in biological computers. As a result, human biology is a computer that will inspire people to choose to develop biology for reasons other than health. Due to the rapid scientific progress with prospective gains in artificial intelligence and genome editing in the coming decades, the answers to all these questions will be more demanding. Not only the Western hemisphere but also the international medical communities are troubled by these apprehensions. In particular, China wants to control artificial intelligence and genome editing internationally, suggesting an effective combination of these techniques in China shortly. We need to balance stimulating inventions, counter social, ethical dilemmas, and ensure responsible development in combined genomics and AI, what shape these developments can take, and how they will be applied to international governance.

Cas9 scissors use advanced technologies to cut DNA in the correct place to be sliced using controlled RNA, involving artificial intelligence during the entire genetic modification [9]. But how the genetic material is sewn together afterward is not very precise; indeed, experts have usually assumed that the process is spontaneous without a template. But "there has been indirect evidence that cells do not spontaneously recover DNA," Richard Sherwood, a geneticist at Brigham and Women's Hospital, said as a researcher [10], and a paper also showed improvements in the 2016 repairs. Sherwood wondered if such results could be predicted by artificial intelligence. Successful sequencing predictions would allow researchers to measure precise guidance on RNAs to induce mutations in humans, providing improved research models for genetic disease study [10]. Sherwood and his collaborators have disclosed their algorithms, which can identify the reference RNA for pathogenic mutations in human patients without a repair template, a clinical application of CRISPR that has been a reality for years, if not decades [10]. The corrections required patients with extraordinary genetic abnormalities, albinism-causing diseases, and blood coagulation failures to function on cell lines, including neural system degradation and growth defects. Market research firm Frost & Sullivan presumed in 2021 that AI would produce a worldwide pharmaceutical income of $6.7 billion [11]. Genomics is the study of an individual's entire set of genes, an area in which artificial intelligence is increasing significantly. While much emphasis has been focused on the risks to human health, gene editing and its use in agriculture and animal husbandry may also be groundbreaking. Once researchers can organize and interpret DNA, which is made more accessible, cheaper, and much more precise by artificial neural networks, they will gain insight into the fundamental genetic strategy that coordinates all events within an organism. With this knowledge, decisions will be made on the cure to which an organism might be susceptible in the future, what mutations may cause different diseases and ways to tackle them.

8.2   GENOME SEQUENCING AND EDITING

Subsequently, our ethnic background has been of great significance to a more in-depth understanding for many decades since people's diseases over a lifetime are mainly guided through their genomes. Our progress was hampered by the importance and complications of data that had to be analyzed. Researchers can deduce and act on data in better ways by using genome sequencing and editing approaches, keeping in mind the ongoing improvements in machine learning and artificial intelligence technologies [12]. The genome sequence is an individual arrangement of the living cells' primary DNA components (adenine, guanine, cytosine, and thymine); the human genome consists of 20,000 gene features, more than 3 billion base pairs. Genome sequencing is a significant initial step towards understanding this. High-throughput sequencing (HTS) is a new technology that allows DNA sequencing to be completed in a single day, which once took a decade to achieve [13]. When DNA changes occur at the mobile level, they are called gene editing. Google's Deep Variant method utilizes the best AI techniques to make high-performance sequences a more accurate representation of a whole genome [14]. While there have been high-throughput sequences available since the 2000s, Deep Variant can distinguish between minor mutations and random errors [14]. Deep learning was essential to Deep Variant's efficient evolution, as shown in Figure 8.2. While we can now quickly recognize and interpret genes, we have recently become oblivious to what they teach us. A Canadian company Deep Genomics utilizes its artificial intelligence program to identify the genome's value when determining the adequate medicinal therapies for a person based on the cell's DNA [15].

The organization's methodology examines mutations and leverages what's discovered in a trove of mutation samples as it's researched to forecast the outcome of a modification. Millions of cancer diagnoses are made each year as a result of documented instances, yet the effectiveness of chemotherapy and treatment is mixed. Doctors will utilise artificial intelligence to classify genetic variants and prescribe

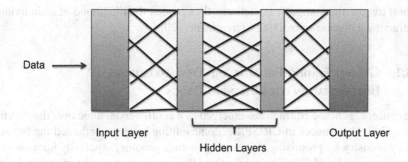

Data ⟶

Input Layer   Output Layer

Hidden Layers

FIGURE 8.2   Architecture of Deep Variant algorithms to handle complex genomic data.

the optimum prescription medicine for each patient, according to service providers like Sophia Genetics [16].

8.2.1   THE GENE-EDITING OPPORTUNITIES AND THREATS

Several other organizations are competing with technologies that enable gene editing by making genetic material changes at the digital level. Data scientists and biologists collaborate with the gene-editing platform CRISPR [17]. The "editing" of genes that can instigate disease or "paste in" genes that produce higher-yielding crops resistant to drought has positive effects. Still, it also has complex social, legal, and ethical implications. By editing mutated genes, many other people can see the advantages of "boosting" health, but as we begin to "optimize" society, the problem is more complicated. Some other problem that researchers are obsessed on is how to prevent outcomes from going beyond the goal when the instruments work on the defective gene, which appears to be the target gene in gene editing. Machine learning and support for artificial intelligence create more convenient, inexpensive, and precise gene editing initiatives. Genetic evaluation tools for infants and children, pharmacogenomics, improvements in agricultural production, and much more are expected to be included in upcoming genetic engineering and artificial intelligence [18]. While it is not possible to predict the future, one thing is sure: machine learning and artificial intelligence will deepen our knowledge of our genetic composition and other life forms as well.

8.3   PERSONALIZED THERAPY AND LIFE-SAVING SERVICES IN THE CONTEXT OF GENOME EDITING

One of the most promising prospects for gene technology is precise or personalized medicine creation [19]. By 2023, it is estimated that the market requiring unique treatments for a patient or group of genetically modified individuals will hit $87 billion. The introduction of personalized medicine [19] has historically been prevented by technology and expense, but machine learning techniques are beginning to overcome these obstacles. Machine learning techniques help identify genetic dataset trends, and computer models can then predict the likelihood of an individual contracting a disease or reacting to treatments [20].

8.3.1 GENOME EDITING INITIATIVE AND PHARMACEUTICALS IMPLEMENTATION FOR DESIGNING DRUGS

The concept "genome editing" has emerged as a controversial topic over the past five years. Current advances in CRISPR genome editing have strengthened the biotechnology industry by promising that any organism's genome, especially humans', can be precisely modified [21]. The productive effect of genome editing in medical management is becoming increasingly evident as many experts are voicing interest in the fundamental science of this technology. Decisions are increasing regarding the clinical use of CRISPR [22]. Still, the importance and significance of this advance in widespread genetic science have begun to be recognized by professionals [23]. The latest strategy used in CRISPR indicates significant improvements over conservative approaches such as effector nucleases (TALEN) and finger nuclease (ZFN). Currently, analysts argue, CRISPR is much more effective, easier to manage, and able to edit multiple genes at the same time compared to conventional methods [24]. In the previous five years, there have been exponential advances in this technology's interest as the first scientists have shown that they can effectively edit DNA independently [25] and the DNA based on organelles [26].

8.4   CRISPR AND GENOME EDITING

CRISPR changes DNA by "deleting" or "pasting" DNA into the template by functioning as a series of "molecular scissors." A short section of the RNA is used in the genomic sequence of interest with the CRISPR-associated (Cas) protein [27]. This purpose makes it simple to use this technology, so experts can only adjust the RNA guide to identify different genes. The Cas complex can modify DNA by using various mechanisms for adding, deleting, or replacing DNA in a CRISPR-Cas system in the target gene [28]. This functionality could prohibit CRISPR-Cas from editing other genes since this can only be corrected when the aim has been met. A set of genetic editing possibilities via using entire CRISPR technology that can be used to improve individual organism potential.

8.4.1 GENOME EDITING FOR THE NEXT GENERATION IN THE CONTEXT OF MEDICINE

As one of the most critical applications for genome editing, next generation medicine(s) will include human medical uses because CRISPR can correct disease-causing mutations [29].. This technology can be used, somatically or with germline alteration, in two different ways by community practitioners. Cognitive therapy refers to DNA editing in a human cell after birth rather than during fetal development [30]. New forms of treatment for emerging diseases, including cancer, will arise using this technology, such as modifying immune cells to treat cancer [31] virtually. The methodology described here is a CRISPR framework, and it is a more contentious gene therapy than previous implementations [32]. CRISPR will occur in gene therapies for a short to a mid-period in healthcare centers, with clinical trials beginning very soon, particularly in the US [33], and more have already started [34] or are about to commence in China [35]. The broader controversy and hope around CRISPR stem from the possibility of potential genomic editing. Genetic disease risk can be avoided by altering the sperm, egg, or embryo genome due to this form of treatment and using in-vitro fertilization to produce a child [36]. Instead of merely treating cancer as for adult patients, it offers the possibility of modifying gene mutations like BRCA1 to reduce the risk of childlike cancer just before birth [37]. Genome editing would also establish that the transformation will have little effect on future generations, leading to deterioration in genetic disorders, particularly Huntington disease or cystic fibrosis [38].

8.5   EPIGENETIC AND GERMLINE EDITING POSITIONS

The germline editing techniques raise both protection and ethical problems for newborn infants and any offspring they produce, as genetic mutations can happen over successive generations [39]. As false target detection happens more often than previously thought, the publication of *Nature Methods* suggests that there may be an increase in the likelihood of alterations in unintended sections of the genome or skewed effects [40]. The active editing is partly performed by embryonic cells in this case, i.e., in a team only [41]. The use of CRISPR germline will lead to mosaicism. These unusual changes can flop and present new health risks in the prevention of ailment. Significant changes in processes involving faster access to CRISPR for the embryo could reduce the off-target impact and epigenetic shift [42]. Several somatic methods cause efficacy problems because the human immune system correctly recognizes the CRISPR-Cas9 enzyme; the enzymes are considered to be foreign, non-human particles, and antimicrobial antibodies are produced to prevent the use of these tools. The clinical application of CRISPR germline editing is most likely to involve Congressional action. Recent assumptions exclude funding any research that could lead to the development or destruction of human embryos, the National Institutes of Health (NIH). They do not endorse the Food and Drug Administration (FDA) to review germline editing products [44]. These barriers scientifically prohibit medical research into the editing of the human germline. While clinical use will need a solution to the above problems, CRISPR germline editing is likely to be clinically available in the intermediate to extended future.

8.6   NATURAL SCIENCE IN THE CONTEXT OF ARTIFICIAL INTELLIGENCE PLATFORM

Evolutions in molecular genetics, biotechnology, and now CRISPR have contributed to biology's conception as a tool. This understanding describes cell and genomic processes using parallels with the concepts of engineering and technologies [45]; mitochondria are classified as "powerhouses" [46], kinases are classified as "moving cargo" motors, and microorganisms producing insulin are classified as factories [48]. Theoretical insights should be expressed primarily utilizing technical analogies that should have second-order consequences such as where and how researchers view academic work and means specific disciplines are evolving. Biology can be viewed as a technology and the philosophy of engineering for the production of modern biological systems for a purpose by examining the underlying principle of synthetic biology [48]. Experts conceive the growth of novel creatures by using the living system's building blocks plus genetic programming capable of supporting a wide variety of tasks [49]. This ability to interface with genomes in living systems enables the editing of genomes in their non-natural biological machines to create new functions [50], which leads to a troublesome approach to living systems. These entities were responsible for the graphical user interface and preserving a DNA record, and researchers were dedicated and committed on the storage of data by genetic makeup [51]. This biological experience in data science represents a broader transformation of the notion of biology as a machine that, like any other system, can be programmed and configured to perform tasks.

8.7 OPTIMIZATION OF HUMAN BIO-MACHINERY

Genome editing may have advanced therapeutic properties, which would open up the "simulation" potential of human biology. Accessing biology as a machine helps conceptualize CRISPR somatic treatments to advance the current paradigm as the equivalent of "pitcher" applications. Nevertheless, CRISPR might enable the editing of human germline embryos and hypothesize the ability of early prototypes or the refining of social, biological tools that are assisted by machines [52]. CRISPR technology has the potential to become commonplace in people's daily lives, thanks to an automated conceptualization of biology among people and health professionals. By inspecting genome editing advances using CRISPR as "patches" or pre-launching therapeutic "debugging" methods, this technique can prove less foreign to end-users.

8.8 GENOMICS IS REVOLUTIONIZED BY ARTIFICIAL INTELLIGENCE

Rational or artificial intelligence (AI) equipment is a sophisticated technology with strong potential to influence this thriving healthcare sector. This tendency has resulted in significant cognitive innovations, allowing AI to gain from its previous experience. This methodology focuses on how young children perceive new knowledge, and the method uses human brain-like machine design [38]. Thus, artificial intelligence recognizes and learns information patterns by analyzing its errors that lead to classifying and categorizing new information. Deep learning effectively allows AI to recognize multiple arrangement layering instruments, such as identifying an organism based mainly on its contours and subsequently acknowledging more explicit details such as wings, pelts, and fur. AI techniques can be applied in several different settings to predict RNA splicing patterns in mouse cells, e.g., identifying animals on Livestream videos [56]. AI technology, which can also lead to new optimization approaches, can also be applied to automation techniques [38]. A strategy can be designed to recognize trends from significant data quantities and optimize structures in association with detailed editing using in-depth capabilities.

This scientific development, rather than its therapeutic applications, can involve adequate studies of the latest useful scientific evidence that may help to construct a prototype of molecular mechanisms, i.e., by what means disease causes genetic makeup variation [57]. Similarly, AI has easily studied vast amounts of genetic datasets from a Google–Genomic England partnership. This ability to better understand how new genetic condition variations can be predicted, detected, distinguished, and discovered will identify AI as a critical clinical pharmacy partner. Cloud computing technology implies further improvement in these technologies, plus patient comfort [58]. AI continues to expand rapidly in the medical industry; AI software has played an enormous role in the clinical use of transmission. The goal of evolutionary research is to adapt corporal devices to serve the desired role; gene editors may use this mindset to improve a patient's capacity to live a healthier life. If human biology is considered a machine, profound learning can be used to educate experts where techniques like CRISPR are applied to optimize the organism through genome management.

8.9 ARTIFICIAL INTELLIGENCE CAPTAINCY AND FRONTIERS FOR THE DIRECT COMPETITORS OF GENOMICS

The use of in-depth research on genomics and gene editing is inconvenient. While AI promises robust new analytical, diagnostic, and systematic approaches in the pharmaceutical field, the latest deep machine learning tools manufacturers don't really adequately explain how modules arrive at their results. [59]. "This technology's mysterious" black box "element can impede risk analysis, such as failure to understand how machine learning" thinks, "and creates problems identifying potential zones in those areas that the system can flop or bug" [60]. When a device makes a diagnostic error that affects the patient and the recovery process of the doctor, understanding why the machine made a mistake may be a challenge. Furthermore, the quantity and existence of scientific information will restrict the accuracy of deep and machine learning algorithms that try to prototype and predict humans' biology [61]. These information limits create a need for time-consuming investigations that confirm their inferences on the gene's medical implications for AI-produced models. The nature plus the amount of information fed into deep learning software limit its potential efficiency. The prevalence and significance for the future of the authentication of medical phenomena predicted by AI may increase, especially as recent facts of the medical implications of pathogenic alterations continue to evolve [62].

8.10   ARTIFICIAL INTELLIGENCE AND GENE EDITING MECHANISMS

Due to the "optimization" of human biology, the concept of spreading AI and AI-based editing will be implemented in the coming years [63]. This section describes whether AI can actively participate in long-term medical decision-making on genetic manipulation, especially given the human genome analysis [64]. Human genome editing will also be used in the near future, as a recent study carried out by the National Academies of Sciences, Engineering, and Medicine suggests embracing this technology under circumstances and diseases without other procedure treatments. This demonstrates where technological research deviate from the norm, necessitating further scientific and ethical scrutiny [65]. In this regard, AI resources can direct clinicians in using recombinant DNA technology in the genome to improve health through the inclusion of gene editing to avoid gene mutation. Deep learning's computing abilities will enable technology to conduct human biology research in the same manner that a high-performance normal computer would do.

AI simulation thus identifies a value-free procedure that does not include clear legal or moral decisions. This could exacerbate controversy over CRISPR's provision of oversized biological capabilities for artificial reproductive improvement. Although the National Academies' new analysis limits the use of technology to cure a disease or improve biology, it does not categorically preclude human enhancement [66]. The academies called for public participation when CRISPR was used to enhance human biological capacities. Still, challenges are posed by identifying improvement and separating it from using through-editing to maximize human health. For example, over appropriate ranges the muscle mass is likely to increase. On the other hand, a patient's significant risk aversion to cardiovascular disease is more difficult to identify because it entails a shift in the entire gene set, as are many other applications of recombinant DNA technology to minimize risk and facilitate welfare [67]. This risk-degrading allele would lead to patients' well-being, and genes could be included in AI methods designed to enhance human health as beneficial recombinant DNA targets. The philosophical challenge of optimizing preventive medicines is possibly not understood by machines as current profound learning techniques work instead of preprogramming by defining designs.

The use of genetics as a tool would obfuscate the distinction between the development of gene editing to better disease control and the spread of genetics as a tool. AI also concluded that the editing of human genomes for better health requires the practice of risk-reducing alleles; agnosticism will express itself against particular improvement issues from the point of view of biology as a device. Since this scientific approach aims to reinforce the target procedure by introducing these genomic changes, it could be the best way to achieve the objective and facilitate the well-being of the patient. Explanations of germline editing may help promote these choices, as changes that enhance well-being would be a viable way of avoiding illness and increasing the child's subsequent autonomy [39]. This viewpoint can also apply to CRISPR use, leading to more noticeable changes if such modifications also give improved health. Similarly, AI has permitted gender-based recombinant DNA applications, raising questions about the utility of distinction.

8.11 SCIENTISTS USED AI TO IMPROVE GENE-EDITING ACCURACY

A substantial majority of prediction models are offered through collaboration between technologists and biologists from research centers across many genome, which increases efficiency and accuracy while using recombinant DNA technology. This gene-editing breakthrough transforms companies from medical coverage to agriculture. The nano-sized CRISPR cutting kit can be programmed at a particular level to slice or modify genetic material within a specific gene. For example, the research will lead to groundbreaking technologies, including particular cell modification to combat chemotherapy or the development of high-yielding and drought-tolerant grains like wheat and maize. Besides, this company uses a branch of AI called machine learning capable of predicting suspected off-target effects when editing genes using the recombinant process. While in many areas, CRISPR shows tremendous potential, a difficulty is that many genomic regions are identical, meaning that the nanoparticle sewing kit could inadvertently begin to work with the faulty gene and cause unintended concerns about the supposed off-target results. "Nicolo Fusi, a scientist at Microsoft's Cambridge, Massachusetts, Research Centre, says," You want to avoid the off-target effects. "You want to make sure your experiment is not destroyed by anything else." In a paper published on January 10, in the journal *Nature*, Fusi and colleagues at MIT Large Institute, Harvard, and University of California, Los Angeles, and former Microsoft colleague Jennifer Listgarten, Harvard Medical School and Massachusetts General Hospital, define accumulation.

8.11.1 GENE EDITING AND BIOMEDICAL ENGINEERING SYSTEMS

Cloud-based all-in-one guide project software works successfully with open source on Microsoft Azure; development and a corresponding software tool known as Azimuth are freely available to predict goal results. The gene of interest has been recognized by researchers, i.e., they want modifications using the vertical components or the cloud-based search tool that enlists scientists' guidance and can filter through predicted impacts on targets and external targets. The gene-editing approach of CRISPR is adapted from a scheme to combat natural viruses. In the late 1980s, scientists found it in bacterial DNA and worked out how it took place over the next few decades. The CRISPR system has not been developed; "it has evolved," said John Doench, a Large Institute associate director who leads the biological portions of Microsoft's research partnership. The quasi-repeat surfaces are fragments of DNA from viral genomes used to classify subsequent viral incursions by molecular markers known as RNA as a guide. The RNA guides the CRISPR complex to the virus upon detecting an invader and dispatches Cas proteins to spur-of-the-moment and deactivate the virus. Throughout 2012, molecular biologists clarified how it is possible to modify the mechanism for battling bacterial viruses to alter organisms' genes ranging from plants to mice and humans. The effect is the manipulating technique of the CRISPR-Cas9 gene. Biologists create synthetic RNA guidance to match a foreign DNA in the gene to those that want to cut or edit or lose it in a cell with the protein cuts associated with CRISPR, Cas9.

The algorithm is currently widely used as an accurate and powerful way to understand the role of local genes in everything from humans to poplar trees and how genes can be changed to do anything from fighting diseases to producing more food. For starters, if you want to learn how gene dysfunction leads to disease, you need to understand first how and why the gene usually works, "Doench said." Here, CRISPR was a real game-changer. In deciding which RNA guide to use for a given experiment, researchers face a global challenge. Each focus is approximately 20 nucleotides; hundreds of potential guides are available for each target gene in a knockout experiment. Generally speaking, each guide's efficacy is different, and the degree of off-target operation also varies. Technologists and archaeologists work together to create instruments that can help researchers navigate the guide's standards and find the best one for their experiments. Several research teams have developed rules for determining where targets are off and how to avoid them in gene editing for any given experiment. "The rules are designed and crafted very much by hand," Fusi said. "We wanted to tackle the problem of machine learning."

8.11.2 TRAINING MODELS FOR GENE EDITING IN THE CONTEXT OF MACHINE LEARNING

To solve the problem, Fusi and Listgarten used data collected by Doench and colleagues to train a machine-learning model known as a first-layer machine-learning model. These data reported activity with just one nucleotide overlap with the guideline for all possible target domains. It incorporates freely publicly accessible data developed by team partners at Harvard Medical School and Massachusetts General Hospital. A second-layer model was fitted by machine-learning specialists who refine and generalize the first-layer model in cases where more than one matched nucleotide is present. The hybrid cryptographic model is critical because the off-target activity can occur with much more than one mismatch between guide and aim, noted Listgarten, who joined the faculty at the University of California, Berkeley, on January 1, 2019. Finally, on several other publicly available datasets, the team validated its two-layer model, and a new dataset was created by associated Harvard Medical School and Massachusetts General Hospital collaborators. Some of the proposed components are intuitive such as, as Listgarten noted, a distinction between the reference sequence and nucleotide. Others reflect new encoded DNA properties, which are discovered through machine learning. "If you give it enough stuff that it can focus on, part of the attraction of machine learning is that it can draw some stuff out," she said.

8.11.3 OFF-TARGET SCORES

Elevation provides researchers with two types of off-target scores per guide: individual scores for one target area and a single summary score for that guide.

Goal-scorers are the machine-learning – the likelihood that anything terrible will happen on the genome for any single region. For any guide, Elevation returns hundreds to thousands of those off-target ratings. Listgarten noted that these particular off-target scores simply could be exhausting for participants attempting to ascertain which of the potentially thousands of guidelines to use for a given experiment. The evaluation score is a specific system lumping off-target scores to provide an overview of how likely the guide is to damage the cell with all of its off-target potentials. "Instead of any point in the genome, what are the chances that I will mess up this cell because of the guide's numerous off-target behaviours?"

8.11.4 END-TO-END GUIDE DESIGN

Writing in *Nature Biomedical Engineering*, the collaborators explain how Elevation works in conjunction with a method they published in 2016 called Azimuth that predicts on-target impact. The complementary tools provide researchers with an end-to-end framework for designing experiments with the CRISPR-Cas9 system, enabling researchers to choose a guide that achieves the intended effect – disabling a gene, such as errors, as cutting the wrong gene.

8.12 EFFICIENT GENOME EDITING AND THE OPTIMIZATION OF HUMANS' HEALTH

Since a computer can be programmed for many tasks, gene editing combined with artificial intelligence can enhance human biology for purposes other than mere health improvement. Recent genetic diagnoses that claim to predict a novice's physical characteristics [70] may be relevant for CRISPR germline editing. Experts refuse to allow knowledge concerning the inherited origins of physical characteristics or personality to become empirically valid [68]. Innovations of the combined overall AI and CRISPR can be seen as having potential for enhancing human biology for physical characteristics and not merely for well-being, thus generating more conventional improvement issues [69]. Discussions on people's preparation for space exploration have also started reasonably [70], including thinking about decreasing height and the body's capacity to respond to radiation. The goal of AI research has been to find more mutants that could improve those properties; genomic analysis already exists to decide if there are spatially useful genetic variants [38]. Analyzing biology as machinery often manipulates these experiences, resulting in initiatives that significantly alter human anatomy to transmit genetic material – a potentially useful characteristic for interstellar living [68]. Engineers should provide the computer with new functionalities, a mindset that expresses itself in addressing how to render humans more capable of photosynthetic usage. Such biological changes would require a great deal of data analysis and optimizations to avoid adverse effects on health, and AI software would undoubtedly support them.

8.13 FURTHER RESEARCH AND DEVELOPMENT FOR GENE EDITING WITH AI

Those technological applications will warrant more advancement in CRISPR integration research and development and in-depth learning because the existing instruments do not carry out these functions. There is still an inadequate explanation of interactions between the variety of genes and diseases because of the sheer number of alleles that continue to be found to have unexplained effects on disease [71]. Therefore, experts who claim to have found reasonably obvious connections between biological variants and diagnostic control have been criticized by the scientific establishment [72]. Researchers have also challenged the importance of using genomics to give potentially excessive advice on preventing diseases and health improvements [73]. This latest branch of the genome industry also directly provides the customer's genetic sequence and offers advice on everything from dietary modifications to better football [74]. Genetic studies that predict facial features in newborns gained even better expert feedback [68]. Recent CRISPR experiments have indicated more difficulties than was initially believed in the inserting of a new gene into a human embryo [75]. The source instead copied DNA from the mother's gene to replace the father's DNA with a new gene, and this could present unique challenges for two copies or improve the progress of diseased genes for the editing of embryos [68, 67].

The "black box" and inadequate AI (quantity or quality) data restrictions also limit the ability to optimize human biology without more computer education work. Progress with deep-learning techniques on their own may be ineffective to increase capacity and to fix AI limitations [76]. The use of artificial intelligence further delays new approaches to technology. It is almost inevitable that sustainable medicines with genetic modification (whether or not with AI) and FDA approval would require more genomics and machine learning experiments, rigorous clinical research, and less restful confusion.

8.13.1 REGULATORY CONSIDERATIONS OF GENE EDITING AND ARTIFICIAL INTELLIGENCE

As artificial intelligence and editing continue to evolve, medical devices that include one or both of these technologies may begin to look for entry into the market. The first diagnostic technology used in deep-learning cardiology clinics was also approved by the FDA earlier this year [77]. Later in 2017, the agency took a final decision on the approval of two therapies to edit the somatic of immune cells and support the newly unanimously recommended Novartis therapy panel [78]. If the authority seeks to understand how and where these tools are to be used in medical equipment, real protocols for future editing and IA changes may also be requested. This is especially important with the application for gene editing since no health tool that modifies the FDA has approved the human gamete in the past and expert information about FDA identification of and bacterial control. If CRISPR gene therapy applications are approved and somatically updated, it is still unclear how the FDA can recognize and control bacteria. Editing of human embryos may be performed differently, while CRISPR control of genetic drugs may be followed by FDA drug regulations for auxiliary reproductive technology [79]. Other problems include the possibility that the FDA will expand the editing unit's control [38].

The AI regulatory framework in genomics increases a further number of risks for the FDA, covered by medical technology regulations and the emerging monitoring community Software-as-a-Medical-Device System (SAMDS). The FDA has expressed interest in developing a digital health system to handle products such as SAMDS, and this goal could be supported by the potential introduction of the new regulations for the 2017 user fee for medical devices [80]. The department has so far refrained from putting artificial intelligence policies in place or talking about them.

Refusing to lay down stringent regulations will prevent legislation from being impenetrable in the industry and potentially burdensome or ineffective rules; however, some have indicated that the FDA is agnostic about "black box" issues from profound medical diagnostic learning [81]. Such problems with the black box may raise unique security and efficiency issues when using CRISPR based on AI and may require further consideration. From a directional point of view, if microbial editing is controlled as a drug, the FDA's auxiliary prognostic analysis can provide a technique to assess the efficacy and safety of genetic research and interpretation of these findings. However, this alternative fails to capture laboratory-developed research, accounting for a large proportion of genomic diagnoses [82]. The FDA also expressed interest in promoting international harmonization of medical device regulatory policy, including the International Forum on Medical Devices Regulators. Such global initiatives will provide opportunities to address ethical and legal jurisdictional dilemmas over issues such as AI-powered editing.

Several other organizations will support successful restrictions on AI-enhanced germline editing. The Centers for Disease Control and Prevention (CDC) has an active office for public health review related to genomics, and genome editing could provide a greater advantage. Using modification approaches, this organisation may be able to aid in the production or assessment of longitudinal health studies of born individuals – a significant step needed for safety monitoring. The Federal Trade Commission (FTC) will play a role in regulating genomics/machine learning technology by using its expertise in educating consumers and correcting false or misleading claims about products. This organization also actively engages in genetic testing (DTC) for the client and will support other regulatory activities around it by editing. Although not a classic regulatory body, by making policy-based decisions about research funding, the NIH will influence the course of through-editing and AI technology. This may include supporting studies to determine the clinical significance of poorly understood genes, evaluating profound learning skills and deficiencies in a clinical setting, and reviewing the impact of the social sciences on AI-based bacterial editing.

8.14   POLICIES AND RECOMMENDATIONS OF GENOME EDITING WITH AI

The US government should commission a study on the ethical, legal, and social implications of using AI to expand medical use. Developing effective control systems depends on analyzing the different potential impacts of integrating these two technologies. This process will be facilitated by gene editing, which will assess consumer attitudes and perceptions of AI-driven possibilities..

The government of the United States supports human genomics and artificial intelligence studies that will promote the risk assessment for one or both strategies, and the abolition of medical methods.

Government regulators, industry representatives, scientists, and other genomics and artificial intelligence specialists should begin discussions on appropriate gamete processing and therapeutic AI monitoring mechanisms, mainly when used together. The key to protecting public health is to find a balance between security and innovation while developing and sustaining US leadership in these areas of emerging technologies. The involvement of the supervisory bodies and their roles and responsibilities of supervision should be considered.

The present government should explore how to collaborate on science and industry initiatives with the biggest economies in both genomics and AI, such as China. International R&D cooperation would encourage more collaborative decision-making on how to conduct ethical research on these technologies, especially in the interface between genomics and in-depth learning.

The policymakers should participate in talks on harmonizing corporate governance for these innovations with leading nations in genomics and AI, both individually and in their convergence, such as China. These discussions involve the Science and Technology Policy Office or another appropriate body to identify and suggest the alignment of various values and standards.

8.15 CONCLUSION

Collaborative work involving technologists and biologists from academic institutions worldwide offers various modeling techniques that improve the efficiency and accuracy of the application of CRISPR, a gene-editing technology that transforms activities from healthcare to agriculture. The modeling method, a study area within quantum computation, is combined with image processing to analyze the enormous proportion of molecular image revisions. Innovative neural networks and recurring neural networks are among the machine learning architectures that now stand out in deep learning. Several machine learning methods address data processing tasks, including some object recognition, binary classification, and information retrieval. Deep understanding is also combined with artificial intelligence to analyze many genomic sequence-related texts found in widely available scientific articles. Deep learning algorithms solve challenges such as detecting specified devices, relational decoding, and data acquisition. Big data strategies can handle in-depth learning projects, as they offer outstanding productivity and address realistic technology challenges.

Innovation in artificial intelligence enables technological development in all development sectors, especially genomics. Reading scientific journals requires patience, and technical language is not easy to understand. That said, at the Virtual University Bioinformatics Expert Laboratory, we are still aware of new biomedical research papers. Twenty cases of machine learning and deep learning tend to be present in genomics. Genome sequencing is genuinely an artificial intelligence field in biological science that focuses on the genome, structure, function, adaptation, simulation, modeling, and editing. Here are some knowledge and imaging applications for genome sequencing: identifying genome and genotypic improvements that are highly productive, assessing genomes, determining genetic substances, predicting protein-protein interactions, and predicting accurate DNA and gene binding receptor sequencing and studies of molecular areas or structures in functional metaheuristic algorithms and how individuals control genetics. Bioinformatics Expert Virtual University Lab explains several frameworks for machine learning and deep learning for practical bioinformatics: the classification of gene expression, prediction of neural activity from genetic composition, prediction of transcription, and prediction of signaling pathways and RNA binding proteins for promoters and enhancers. Operational bioinformatics, the field of molecular genetics that aims to clarify genomic knowledge and relationships, is computational biology. Here are several AI solutions for systems biology: the classification of variations and metabolic events, classification of expression patterns, prediction of sequence regulators and stimulators, and prediction of gene transcription and RNA-containing receptors. Systemic therapy: systemic evolutionary computing in the field of nanotechnology that requires the analysis of molecular configurations. In the Bioinformatics Expert Virtual University Lab, we identify several dimensions of prototype artificial intelligence and machine learning: the classification of proteins in the active site, type of protein complexes, prediction of interaction diagrams, prediction of molecular structure, and prognosis of secondarily associated proteins.

Concepts for AI genomics: users participate in machine learning algorithms, but they do not know how to get started. At the Bioinformatics Expert Virtual University Lab, developers began exploring several methods for applying artificial intelligence in genomics. Here are 35 suggestions for AI genomics: editing configuration and chromosome-based mutations, diagnosing cancer regulatory trends, distinguishing genetic mutations, sorting genomic profiling, categorizing mutant types, designing selective therapies, revealing deoxyribonucleic transcriptome genomic domains drug therapies, cancer and cancer differences, identifying and distinguishing receptor binding sites, recognizing transcription factor binding sites, splice sites, exons, interpreting single-cell regulatory control, mode long regulatory components, differentiating and more by chromatin status, predicting genomic DNA sequence chromosome tags or diagnosing disease phenotype, predicting genetic expression, identifying genomic interactions, defining amino acid sequence polymer compounds, predicting regulatory mechanisms and conditions, estimating domain compatibility of amplification regions, and estimating and bees da ribonucleic acid receptors, identifying the splicing behavior of specific exons, identifying variant lesions, and characterizing the effects of particular nucleotide mutations on the availability of chromatic acid.

Several issues require the use of machine learning to solve biochemical problems. These legal issues now give artificial intelligence providers, such as the Bioinformatics Specialist Virtual University Laboratory, the opportunity to overcome competitive challenges and develop advanced analytics. Here are three genomic possibilities: constructing simple truth identifiers or molecular historical data can be expensive. It is essential that the concepts of "freedom to explain" are overcome, and studies are allowed. Wellness configuration is probably the first way these tools can be used effectively. But it may also be possible to optimize basic biology for other aspects, such as physiological or potentially neurological characteristics. Besides, it will enable humans to optimize for different tasks, such as space technology. For example, the use of AI and CRISPR for simulation would require extensive approval. New studies suggest that although this is not always the case, the general public is not prepared to start manipulating DNA in human cells. Thinking of "technology as a machine" is a way of counteracting public sentiment for calculation. Experts tend to talk about scientific information in terms often used in computers and use DNA as "software" for this computer device, such as "business cells." If the world begins to know about gene editing as "updating" or "regression," the concept may be less foreign and much more appropriate. Moral, legal, and social issues are not limited to the Western world. The study describes several individuals who study genomics and AI. Investing heavily in both domains, the Chinese government is an important participant. Different countries may acquire the ability to optimize biology and may have contrasting support for optimization from their entities. International authorities must also think carefully about preventing problems that may come from different countries with different AI-driven CRISPR editing strategies.

Current research based on collected data still shows significant results of gene-editing systems for manipulating treatment strategies for various biological disorders, among which the CRISPR-Cas9 method has so far been incredibly effective, when establishing cross-functional techniques, or indirectly stimulating genetic positions. In the future, a synthesis of integrated CRISPR sequencing and existing data on the genetic or epigenetic properties of carcinoma cells will be able to perceive compounds of intramolecular charge in the genome to a large extent and facilitate the production of new chemotherapy drugs. The CRISPR-Cas9 platform also offers a new tool for changing non-coding regions in the cancer genome, speeding up the functional exploration of aspects that have so far been poorly represented. The enormous development in the treatment of nuclease technology (especially ZFN, TALEN, and CRISPR-Cas9) laid the foundation for the systematic editing of clinical science's theoretical structure. The direction of ZFN via an AAV for in vivo gene editing was the world's first source to describe the diagnosis of genetic defects by in vivo gene editing, which further showed that gene editing has significant potential for clinical use in congenital disability diagnosis. At the same time, gene-editing methods have also enabled bio sensing, gene regulatory control, gene expression modification, the development of chemotherapeutic agents, genomic penetration, and genetic diagnostics.

Although more simulations is considered necessary to account for the off-target effect in the integration of gene modification, detailed integrated editing structures and more oriented nanocomposite bodies have embraced technology and reduced toxicity during the implementation phase, introducing the entire genome editing technology to the treatment center. With further exploration of this technology and collaboration between the global scientific communities, it is possible to believe that the technology can eventually illuminate the biological processes behind disease formation and progression through editing, providing new treatments, and ultimately encouraging advances in life sciences.

8.16   FUTURE PROSPECTS OF AI

AI and genome editing are aimed to include gene editing, stem cell therapy options for embryogenesis, breakthroughs in domestic animals, and more. Although the future cannot be predicted, one thing is sure: artificial intelligence and machine learning can accelerate our understanding of our biological makeup and other natural systems. The toolbox for full genome editing is an effective plan for teamwork to eliminate harmful diseases. Such GE methods can play an essential role in ensuring the biological safety of various micronutrients by modifying genomes that confer resistance to the microorganism. To assess certain throughput impurities, related frameworks can also be incorporated. It is also possible to detect a beneficial response to the presence of harmful microorganisms, which improves the quality of high-contrast growth. This can also provide details about the biological mechanisms of a viral infection or the anatomy and physiology of a microscopic organism by explicitly removing or replacing various genes involved in initiation and progression. GE can also clarify the understanding of key actors that generate switch-pathogen interactions, and the role of different signaling molecules and receptor proteins for possible gene expression. It is also possible to target epigenomes and abiotic stresses, to increase physiological function in humans. Off-target effects and the uncontrolled introduction of expression vectors into the DNA insertion genome reduce their broader consequences. The safe inclusion of plasmid-mediated RGENs in the target gene can lead to adverse effects and immunogenicity.

Advances in gene editing, such as creating wire exchange methods for Cas9 RNP or donor DNA for genetic transformation for gene editing, can be successfully used to generate predominantly mutants without promoting genomic vector intervals. For the first generations, however, a careful analysis of suitably modified species must be performed using a variety regularly to ensure that mutations are stable and rule out the negative effect of gene editing on growth and development. To broaden the awareness of critical ethical issues, their theoretical significance and incorporation must be communicated through users and resource groups. Thanks to its enormous potential to make desired changes in the genome and flexible therapeutic studies, CRISPR/Cas-based genome editing has developed as a significant opportunity in recent history. Scientific innovations are now applied to produce knockdowns and produce throws and modulation and manipulation of gene regulation. Many developments have already been made in a short time since its discovery due to its usefulness in a applied method, including DNA-free through-editing systems (RNPs), many Cas9 versions, multiple multi-gene strategies, precise base editing, and steps to increase the frequency of HDR. The extensive use of this powerful through-editing method has also led to advances in CRISPR/Cas-related bioinformatics tools. However, it is not possible to discount the task of having a society with appropriate information about the CRISPR/Cas-mediated organism modification method, as this is crucial to get laboratory research to the masses.

In one case, it is necessary to have rules and regulations that specifically distinguish between organisms genetically modified with foreign DNA and non-foreign DNA, which means that all legislation will circumvent the latter, allowing very trouble-free use of such organisms. Compared to the commercial release of transgene-free genome-edited microorganisms that have a newly mutated gene, especially in countries where policies are not in place, genome-edited plants with the same mutation occur with natural variation will be more relaxed. In this respect, careful editing methods seem promising. The difference between the research facility and the field is continually increasing due to legislative measures that are not yet well identified. While gene editing can lead to naturally occurring genetic modifications, the effects of genetic changes applied to "modern" technology and editing cannot be distinguished under certain circumstances. Therefore special technical regulation is not rational. Another view is that the representative should focus on the finished product's essential characteristics rather than the production process to identify the potential threat in any case. At the same time, it is not possible to disregard the responsibility to carry out an adequate risk assessment, to monitor the genetic motivation of the substance, and to take appropriate measures to adopt this technology to reproductive and hereditary traits. In addition to these concerns, new measures are needed to develop transformation techniques that are not yet standardized for different species of organisms and stand out as a bottleneck to virtually explore the latest advances. CRISPR/Cas-based through-editing is still limited to a complex molecular biology laboratory that focuses primarily on biological processes.

On the contrary, in crop development projects, raw material producers still need to find good ways to integrate technological progress. Foreign and domestic compensation use facilities that can provide the resources to create, transform, and evaluate carefully edited organisms will be a game-changer when it comes to moving from field to field by offering the most potent through-editing tools for growers. The information here will help us to understand the latest technological advances, research gaps, and practical integration issues to allow us to successfully use organism editing tools.

BIBLIOGRAPHY

1. Doudna, J. A. (2020). The promise and challenge of therapeutic genome editing. *Nature, 578*(7794), 229–236.
2. Maule, G. (2020). Genome editing strategies to restore altered splicing events.
3. Broeders, M., Herrero-Hernandez, P., Ernst, M. P., van der Ploeg, A. T., & Pijnappel, W. P. (2020). Sharpening the molecular scissors: advances in gene-editing technology. *Iscience, 23*(1), 100789.
4. Gullapalli, R. R. (2020). Evaluation of commercial next-generation sequencing bioinformatics software solutions. *The Journal of Molecular Diagnostics, 22*(2), 147–158.
5. Lin, X., Chemparathy, A., La Russa, M., Daley, T., & Qi, L. S. (2020). Computational methods for analysis of large-scale CRISPR screens. *Annual Review of Biomedical Data Science, 3*, 137–162.

6. Tong, Y., Whitford, C. M., Blin, K., Jørgensen, T. S., Weber, T., & Lee, S. Y. (2020). CRISPR–Cas9, CRISPRi and CRISPR-BEST-mediated genetic manipulation in streptomycetes. *Nature Protocols*, *15*(8), 2470–2502.

7. Li, H., Yang, Y., Hong, W., Huang, M., Wu, M., & Zhao, X. (2020). Applications of genome editing technology in the targeted therapy of human diseases: mechanisms, advances and prospects. *Signal Transduction and Targeted Therapy*, *5*(1), 1–23.

8. Dzobo, K., Adotey, S., Thomford, N. E., & Dzobo, W. (2020). Integrating artificial and human intelligence: a partnership for responsible innovation in biomedical engineering and medicine. *OMICS: A Journal of Integrative Biology*, *24*(5), 247–263.

9. Mohamadi, S., Zaker Bostanabad, S., & Mirnejad, R. (2020). CRISPR arrays: a review on its mechanism. *Journal of Applied Biotechnology Reports*, *7*(2), 81–86.

10. Hampton, T. (2020). Virus surveillance and diagnosis with a CRISPR-based platform. *JAMA*, *324*(5), 430–430.

11. Sharma, U., Tomar, P., Bhardwaj, H., & Sakalle, A. (2020), artificial intelligence and its implications in education. In *Impact of AI Technologies on Teaching, Learning, and Research in Higher Education* (pp. 222–235). IGI Global.

12. Ilyas, M., Mir, A., Efthymiou, S., & Houlden, H. (2020). The genetics of intellectual disability: advancing technology and gene editing. *F1000Research*, *9*.

13. Dritsoulas, A., Campos-Herrera, R., Blanco-Pérez, R., & Duncan, L. W. (2020). Comparing high throughput sequencing and real time qPCR for characterizing entomopathogenic nematode biogeography. *Soil Biology and Biochemistry*, 107793.

14. Huang, P. J., Chang, J. H., Lin, H. H., Li, Y. X., Lee, C. C., Su, C. T., ... & Chiu, C. H. (2020). Deep variant-on-spark: small-scale genome analysis using a cloud-based computing framework. *Computational and Mathematical Methods in Medicine*, *2020*.

15. Koumakis, L. (2020). Deep learning models in genomics; are we there yet?. *Computational and Structural Biotechnology Journal*.

16. Mishra, R., & Li, B. (2020). The application of artificial intelligence in the genetic study of Alzheimer's disease. *Aging and disease*, 0.

17. Stangl, C., de Blank, S., Renkens, I., Westera, L., Verbeek, T., Valle-Inclan, J. E., ... & Voest, E. E. (2020). Partner independent fusion gene detection by multiplexed CRISPR-Cas9 enrichment and long read nanopore sequencing. *Nature Communications*, *11*(1), 1–14.

18. Aiman, S., & Patil, K. K. (2020). From generic to custom: a survey on role of machine learning in pharmacogenomics, its applications and challenges. In *Evolution in Computational Intelligence* (pp. 33–42). Springer, Singapore.

19. De Rose, D. U., Cairoli, S., Dionisi, M., Santisi, A., Massenzi, L., Goffredo, B. M., ... & Auriti, C. (2020). Therapeutic drug monitoring is a feasible tool to personalize drug administration in neonates using new techniques: an overview on the pharmacokinetics and pharmacodynamics in neonatal age. *International Journal of Molecular Sciences*, *21*(16), 5898.

20. Abdulqader, D. M., Abdulazeez, A. M., & Zeebaree, D. Q. (2020). Machine learning supervised algorithms of gene selection: a review. *Machine Learning*, *62*(03).

21. Ledford, H. (2015). CRISPR, the disruptor. *Nature News*, *522*(7554), 20.

22. Zhang, Y., & Karakikes, I. (2020). Translating genomic insights into cardiovascular medicines: opportunities and challenges of CRISPR-Cas9. *Trends in Cardiovascular Medicine*.

23. Hsu, P. D., Lander, E. S., & Zhang, F. (2014). Development and applications of CRISPR-Cas9 for genome engineering. *Cell*, *157*(6), 1262–1278.

24. Nemudryi, A. A., Valetdinova, K. R., Medvedev, S. P., & Zakian, S. M. (2014). TALEN and CRISPR/Cas genome editing systems: tools of discovery. *Acta Naturae (англоязычная версия)*, *6*(3 (22)).

25. Jinek, M., Chylinski, K., Fonfara, I., Hauer, M., Doudna, J. A., & Charpentier, E. (2012). A programmable dual-RNA–guided DNA endonuclease in adaptive bacterial immunity. *Science, 337*(6096), 816–821.

26. Cong, L., Ran, F. A., Cox, D., Lin, S., Barretto, R., Habib, N., ... & Zhang, F. (2013). Multiplex genome engineering using CRISPR/Cas systems. *Science, 339*(6121), 819–823.

27. Barrangou, R., & Doudna, J. A. (2016). Applications of CRISPR technologies in research and beyond. *Nature Biotechnology, 34*(9), 933–941.

28. Kim, H., & Kim, J. S. (2014). A guide to genome engineering with programmable nucleases. *Nature Reviews Genetics, 15*(5), 321–334.

29. Baltimore, D., Berg, P., Botchan, M., Carroll, D., Charo, R. A., Church, G., ... & Greely, H. T. (2015). A prudent path forward for genomic engineering and germline gene modification. *Science, 348*(6230), 36–38.

30. Regalado, A. (2015). Engineering the Perfect Baby. Scientists are developing ways to edit the DNA of tomorrow's children. Should they stop before it's too late.

31. Begley, S. (2016). Federal panel approves first use of CRISPR in humans. *STAT.*

32. Heidari, R., Shaw, D. M., & Elger, B. S. (2017). CRISPR and the rebirth of synthetic biology. *Science and Engineering Ethics, 23*(2), 351–363.

33. Begley, S. (2016). First CRISPR trial in humans is reported to start next month. *STAT.*

34. Rana, P. (2017). China pushes ahead with human gene-editing trials. *Wall Street Journal.*

35. LePage, M. (2017). Boom in human gene editing as 20 CRISPR trials gear up. *New Scientist.*

36. Regalado, A. (2015). Industry body calls for gene-editing moratorium.

37. Corbyn, Z. (2015). Crispr: is it a good idea to upgrade? *The Observer.*

38. Johnson, W., & Pauwels, E. (2017). How to optimize human biology.

39. Christopher, G., Thomas, D., & Julian, S. (2017). The ethics of germline gene editing.

40. Schaefer, K. A., Wu, W. H., Colgan, D. F., Tsang, S. H., Bassuk, A. G., & Mahajan, V. B. (2018). Retraction: unexpected mutations after CRISPR–Cas9 editing in vivo.

41. Le Page, M. (2017). Mosaic problem stands in the way of gene editing embryos. *New Scientist, 3117.*

42. Marcus, A. D. (2017). In gene-editing advance, scientists correct defect in human embryos. *Wall Street Journal.*

43. Reik, W. (2007). Stability and flexibility of epigenetic gene regulation in mammalian development. *Nature, 447*(7143), 425–432.

44. Cook, P. Consolidated Appropriations Act, 2017, Pub. L. No. Pub. L. 115–31, 1 (2017).

45. Pauwels, E. (2013). Communication: mind the metaphor. *Nature, 500*(7464), 523.

46. Javadov, S., & Kuznetsov, A. V. (2013). Mitochondria: the cell powerhouse and nexus of stress. *Frontiers in Physiology, 4,* 207.

47. Kevenaar, J. T., Bianchi, S., Van Spronsen, M., Olieric, N., Lipka, J., Frias, C. P., ... & Hilbert, M. (2016). Kinesin-binding protein controls microtubule dynamics and cargo trafficking by regulating kinesin motor activity. *Current Biology, 26*(7), 849–861.

48. Serrano, L. (2007). Synthetic biology: promises and challenges. *Molecular Systems Biology, 3*(1), 158.

49. Delgado, A. (2016). Assembling desires: synthetic biology and the wish to act at a distant time. *Environment and Planning D: Society and Space, 34*(5), 914–934.

50. MacDonald, I. C., & Deans, T. L. (2016). Tools and applications in synthetic biology. *Advanced Drug Delivery Reviews, 105,* 20–34.

51. Erlich, Y., & Zielinski, D. (2017). DNA fountain enables a robust and efficient storage architecture. *Science, 355*(6328), 950–954.

52. Million-Weaver, S. (2016). *UW Engineers, Physician Work Together to Develop Advanced Cancer Therapies*. UW School of Medicine and Public Health.

53. LeCun, Y., Bengio, Y., & Hinton, G. (2015). Deep learning. *Nature*, *521*(7553), 436–444.

54. Knight, W. (2017). Is Artificial Intelligence Stuck in A Rut?

55. Oremus, W., & Land, M. (2012). In Artificial Intelligence Breakthrough, Google Computers Teach Themselves to Spot Cats on YouTube.

56. Leung, M. K., Xiong, H. Y., Lee, L. J., & Frey, B. J. (2014). Deep learning of the tissue-regulated splicing code. *Bioinformatics*, *30*(12), i121–i129.

57. Kim, T. H. (2017). A Review of deep genomics applying machine learning in genomic medicine. *Hanyang Medical Reviews*, *37*(2), 93–98.

58. Sharma, R. (2017). Cloud Computing Companies Move into Medical Diagnosis.

59. Castelvecchi, D. (2016). Can we open the black box of AI?. *Nature News*, *538*(7623), 20.

60. Knight, W. (2017). The Dark Secret at the Heart of AI'11 April 2017.

61. Gluckman, P. D., Hanson, M. A., & Low, F. M. (2011). The role of developmental plasticity and epigenetics in human health. *Birth Defects Research Part C: Embryo Today: Reviews*, *93*(1), 12–18.

62. Begley, S. (2017). Genome sequencing raises alarms while offering patients few benefits. *STAT*.

63. Marriott, N. (2017). SOPHiA artificial intelligence empowers liquid biopsies to fight cancer. *Drug Target* Review.

64. Stockton, N. (2017). Veritas genetics scoops up an AI company to sort out its DNA. Electronic resource]–Mode of access: https://www.wired.com/story/veritas-genomics-scoops-up-an-ai-company-to-sortout-its-dna/–Title from the screen.

65. Nuffield Council on Bioethics. (2016). *Genome Editing: An Ethical Review-a Short Guide*. Nuffield Council on Bioethics.

66. Ormond, K. E., Mortlock, D. P., Scholes, D. T., Bombard, Y., Brody, L. C., Faucett, W. A., ... & Musunuru, K. (2017). Human germline genome editing. *The American Journal of Human Genetics*, *101*(2), 167–176.

67. Maron, D. F. (2015). Improving'humans with customized genes sparks debate among scientists. *Scientific American*, *3*, 2015.

68. Regalado, A. (2017). Upfront Baby Genome Sequencing for sale in china.

69. Long, H. (2013). 'Designer babies': the ultimate privileged elite?. *The Guardian*.

70. Regalado, A. (2017). Engineering the Perfect Astronaut.

71. Cheon, J. Y., Mozersky, J., & Cook-Deegan, R. (2014). Variants of uncertain significance in BRCA: a harbinger of ethical and policy issues to come?. *Genome Medicine*, *6*(12), 121.

72. Howrigan, D. (2013). Guest post: the perils of genetic risk prediction in autism. *Genomes Unzipped*.

73. Brown, K. V. (2017). Scientists Push Back Against Booming Genetic Pseudoscience Market.

74. Bohannon, J. (2017). The cyberscientist.

75. Ma, H., Marti-Gutierrez, N., Park, S. W., Wu, J., Lee, Y., Suzuki, K., ... & Darby, H. (2017). Correction of a pathogenic gene mutation in human embryos. *Nature*, *548*(7668), 413–419.

76. Knight, W. (2017). The dark secret at the heart of al. *Technology Review*, *120*(3), 54–61.

77. Marr, B. (2017). First FDA approval for clinical cloud-based deep learning in healthcare. *Forbes*, Jan, 20.

78. Feuerstein, A., & Garde, D. (2017). Novartis CAR-T cancer therapy wins expert support for FDA approval. *STAT*.

79. Grant, E. V. (2016). FDA regulation of clinical applications of CRISPR-CAS gene-editing technology. *Food & Drug LJ, 71*, 608.
80. Hsu, J. (2017). FDA assembles team to oversee AI revolution in health. *IEEE Spectrum: Technology, Engineering, and Science News.*
81. Brouillette, M. (2017). Deep Learning Is a Black Box, but Health Care Won't Mind.
82. Javitt, G. H. (2007). In search of a coherent framework: options for FDA oversight of genetic tests. *Food & Drug LJ, 62*, 617.

9 Artificial Neural Network (ANN) Techniques in Solving the Protein Folding Problem

Raghunath Satpathy

CONTENTS

9.1 INTRODUCTION

All functional proteins adopt a unique compact three-dimensional (3D) conformation from their unfolded amino acid sequences. Protein folding is a phenomena by which the proteins assembles itself into their correct native structure by orientating and arranging the amino acids relative to each other.. (Figure 9.1). There are problems associated with the folding process, such as how the folding occurs, what the pathways are, what imparts overall stability after folding is over, and overall the basic rules for the protein folding, etc. So all these queries are known as the *protein-folding problem* [1–2].

The historical work regarding protein folding started with the famous experiment by Anfinsen in the year 1961on the enzyme ribonuclease A. As per Anfinsen's observation, the ribonuclease enzyme can fold spontaneously into its native stable

DOI: 10.1201/9781003126164-9

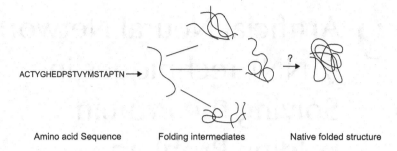

ACTYGHEDPSTVYMSTAPTN →

Amino acid Sequence   Folding intermediates   Native folded structure

FIGURE 9.1   A rough demonstration of the protein folding problem.

state conformation by the pairing of cysteine residues to form disulfide bonds, and this process is reversible. He also observed that the incorrect disulfide bonds that are formed could be reduced and rearranged with the help of a reducing agent [3]. The works of Anfinsen opened the door to understanding the fundamental phenomena that the translation of mRNA produces an unfolded form of the peptide, and it gradually folds progressively to achieve its stable conformation, thereby forming a functional protein. In the year 1968, Levinthal pointed out the existence of a large number of degrees of freedom in an unfolded protein and the possible conformation, if computed for the molecule, may reach an astronomical number. For example, if a given protein has 100 amino acid residues, and if only 3 possible configurations for each residue are assumed, there will be about 3^{100} possible conformations. If 10^{13} s is required to convert one form of conformation into another, then the search for all conformations for the 100 amino acid residues protein will require 10^{27} years, which is quite a long time. Most of the protein follows a specific pathway from the unfolded conformations to their native state conformations. So it is interesting to understand the correct folding pattern and pathway of the protein [4]. Later on, the study of the protein folding problem generated many models by researchers to explain the protein folding process (Table 9.1).

9.2 ROLE OF MOLTEN GLOBULES IN PROTEIN FOLDING

As per Levinthal's explanation, the possible conformations for a protein are very numerous and often follow a particular folding pathway in the protein folding process. More specifically, the protein has to undergo an intermediate structure (like the compact secondary structure) and then acquire the stable 3D structure in the native state. In the case of globular proteins, the presence of these unfolding intermediates has been reported [9, 10]. Ohgushi and Wada and Dolgikh et al. termed such structure as *molten globules* [11, 12]. Kuwajima et al. and Ikeguchi et al. studied the molten globule state in the case of alpha-lactalbumin protein, which was identical to its folding intermediate [13, 14]. Later on, it was proposed that the molten globule state is a common intermediate during the protein folding process and thus considered as a critical factor for the experimental protein folding studies [15, 16, 17].

TABLE 9.1

Different Models Framed on Protein Folding

S.N.	Name of the Model	Overview of Postulates of the Model	References
1	Framework model	Secondary structures are independent of the three-dimensional structure upon folding process	[5]
2	Hydrophobic collapse model	Protein collapses rapidly around the hydrophobic side chains and gradually rearranges to form the three-dimensional structure	[6]
3	Nucleation model	The neighboring residues of the protein interact to form a secondary structure, and that gradually forms a nucleus around which a three-dimensional structure is made	[7]
4	Jigsaw model	Each molecule folds by taking a specific pathway during the folding process	[8]

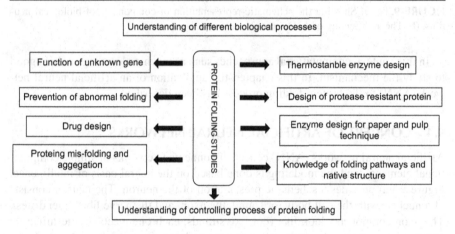

FIGURE 9.2 Importance of protein folding study.

9.3 IMPORTANCE OF PROTEIN FOLDING STUDIES

The current research focus aims to uncover the complex principles that govern the protein folding process by using both experimental and theoretical approaches. However, the detailed mechanism of folding is not entirely known yet. There is the significant importance of the application of ANN to study and predict the protein folding process that would lead to many practical applications , starting from protein function prediction to drug design (Figure 9.2). The in vivo protein folding process within the cell differs significantly from the in vitro condition. The polypeptides newly formed by the cell initiate the folding process even before its complete synthesis. Although there are tremendous biotechnological applications in the study of the protein folding process (Figure 9.3), complex and sophisticated instruments are required to study the mechanism [18, 19].

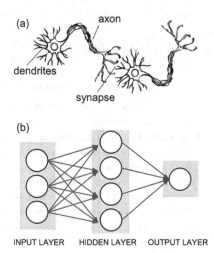

FIGURE 9.3 (a) Showing the schematic representation of components of biological neurons. (b) The architecture of ANN contains different layers.

In this context, computational approaches such as machine learning are held good to study the mechanism. In this chapter, the application of an artificial neural network (ANN) to the protein folding process has been described.

9.4 CONCEPT OF ARTIFICIAL NEURAL NETWORKS

Artificial neural networks (ANN) consist of connected networks of simple computational elements, and the modeling is done based on the neural basis of intelligence. Figure 9.3(a) provides a schematic presentation of the neuron. The neuron consists of a nucleus with the cell body (or soma), the axon, and the nerve fibers (dendrites). The axon components look like root-like strands; each one meets at a terminating synapse junction on a dendrite (cell body) of another neuron.

ANN functions analogously to the biological brain and performs functions of the biological brain. The basic concept used to develop the mathematical model is the crucial understanding of the neuron's biological functioning and the way and the pattern of interconnections among them. In Figure 9.3(b), the common mode of arrangements of all network architectures is shown. The input layer is the outermost one and it receives the input signal (parameters) and the output layer generates the result. In addition to this the third layer is known as the hidden layer. All the node values from the input layer are multiplied and further summed up to calculate a value in the hidden layer of the network that provides the input for the next output layer.

Typically, the ANN is developed on some of the important characteristics of neurological activity. The basic model of ANN is presented in Figure 9.4. Each input entity is multiplied by its corresponding weight. Finally, the weighted inputs are summed up to determine the activation level by implementing an activation function. The input vector $X = (x^1, x^2, x^3..., x^n)$ corresponds to the signals into the synapses

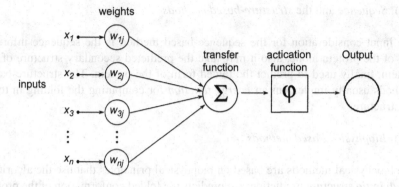

FIGURE 9.4 Showing processing of the neural network.

of a biological neuron and the weights = (w_{1j}, w_{2j}, w_{3j}, ..., w_{nj}) that correspond to the strength of the biological connection. The sum of weighted input is further processed by an activation function (logistic, hyperbolic, or linear type) to produce the activated output signal. ANN models are specified by the net topology and characterized by the training or learning (Figure 9.4) [20, 21].

The ANN training process is performed by applying the data supplied by the input vector form while adjustment of the network weights is done as per the required algorithm. During training, the network weights are gradually converted to each input's values that produce the desired output. The learning process in the case of ANN requires adjusting the input network of weights to produce the output. The training procedure consists of two types, *supervised training* and *unsupervised training*. In the case of a supervised training process, an input vector and a target vector representing the desired output are computed by comparing the corresponding target vector. The error is minimized by the adjustment of the change in the weight of the network. Unsupervised learning requires only the input vector as the training set, and the network weight is adjusted as per the training algorithm. There are three types of ANN structures used to develop the model: feed-forward networks, feedback networks, and self-organizing networks. The feed-forward networks usually cause the transformation of a set of input properties into sets of output signals by adjusting the system parameters. Similarly, in the case of self-organizing network, unsupervised form of ANN is used for pattern recognition [22, 23].

9.5 CRITERIA AND EVALUATION OF APPLICATIONS OF ANN IN THE PROTEIN FOLDING PROBLEM

The computational approaches to the protein folding problem can be focused on. For a given the amino acid sequences (primary structure) of a protein, the problem is to find the three-dimensional (folded) structure. Several methods have been formulated to study protein folding. These methods can be further classified into two categories based on the approaches they follow [24, 25]:

(a) *Sequence* and the *structure-based methods*

The input consideration for the sequence-based method is the sequence informa-tion of the protein alone; also it may take the predicted secondary structure of the protein, finally used to predict the folded form of the protein. The structure-based methods usually implement an *energy function* for computing the folding in to its 3D structure.

(b) *Biophysics-based methods*

The biophysical methods are based on biophysical principles that use the algorithm for *ab initio* structure prediction that predicts the folded conformation of the protein staying at the global free energy minimum state.

9.6 BIO-INSPIRED OPTIMIZATION ALGORITHMS THAT CAN BE USED FOR PROTEIN FOLDING STUDY IN ASSOCIATION WITH ANN

Bio-inspired algorithms implement *natural process*-derived ideas for model devel-opment. This key principle of these methods is *natural evolution*. In addition to this, the behaviors of the different organisms are also taken into consideration. The basis for the development of such algorithms is the use of random search methods and heuristic principles to find the problem's optimal solution. The bio-inspired methods possess many significant features and are advantageous for their suitable implementation for the optimization algorithm [26, 27]. These criteria can be used to evaluate the performance of ANNs to solve the protein folding problem. In addi-tion to that, relevant input data and training algorithms are the crucial consider-ations to minimize the error level and to enhance the prediction accuracy. Also, it is desirable to compare the accuracy of the different algorithms to generalize the rule for the mechanism of prediction. Current trends in machine learning employ hybrid algorithms consisting of the different bio-inspired algorithms and the ANN to form novel methods with good prediction accuracy to solve the protein folding problem (Figure 9.5).

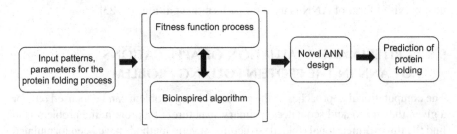

FIGURE 9.5 Architecture of ANN with the bio-inspired algorithm for the protein folding process.

TABLE 9.2
Major Bio-Inspired Algorithms Used for the Study of the Protein Folding Process

S.N.	Name of the Algorithm	Algorithm Basis	References
1	Genetic algorithm	Genetic algorithm (GA) is based on biologically inspired principles of genetics. Initially, some of the best conformations of protein are taken, and some of these conformations are applied mutation and crossover operations. If the new conformation energy is lower than previous ones, then the population is updated with this new one.	[30]
2	Ant colony system approach	It uses a colony of artificial ants allowed to search the possible pathways (optimum solutions). In this case, the problem is represented as the graph and the ants walk on the graph to predict solutions.	[31]
3	Evolutionary Monte Carlo simulations	This method combines the features of a genetic algorithm (GA) and Monte Carlo sampling techniques. This also uses the principle of GA, such as mutation, exchange, and crossover operators, to speed up the search. Typical steps in this method include the following. 1. Make a random move to produce a new conformation. 2. Calculate the energy change for the new conformation. 3. Accept or reject the move based on the criterion.	[32]
4	PFOLD-P-ACO protein folding algorithm	In this approach a hybrid and heuristic approach is used to solve the protein folding problem in the shortest possible time frame.	[33]
5	Branch and bound algorithm	This algorithm is based on chain growth in which the monomers are added one by one, and the energy of partial conformation is evaluated. If there is overall improvement in the energy of partial conformation, then that movement in that direction is considered good.	[34]

Previous studies proposed several bio-inspired algorithms to optimize ANN to improve accuracy and speed [28]. The nature-inspired algorithms have shown their effectiveness and generate the optimal ANN parameters [29]. However, most of the parameters influences the evolution of the synaptic weights, ,the number of the hidden layer and the evolution of transfer functions, and so on (Table 9.2).

9.7 IMPLEMENTATION OF ARTIFICIAL NEURAL NETWORK METHODS IN PROTEIN FOLDING STUDIES

Lately, the artificial neural networks have been widely used for the secondary and tertiary structure prediction of the proteins by considering the amino acid sequence data. Due to the current advancements in machine learning methods, many scientists have applied these particular methods to solve the problem of protein folding. Also, in the past few years, many sophisticated methods have been developed by researchers to find the correlation between the protein sequences and structures as well as the folding rates. Bohr et al., in 1990, described the method for the three-dimensional structure prediction of protein backbones by using neural networks. In this method, functional homologous proteins were used as the training set by using the feed-forward neural network using the back propagation algorithm. The basis for the 3D structure of the protein was to take the *binary distance constraints* for the Cα atoms present in the protein backbone. Further the steepest descent minimization approach was utilized along with the trained neural network to predict the folded 3D conformation of the protein [35].

Similarly, the back propagation neural network has been designed by Sacile, R., and Ruggiero to establish the correspondence between the tensional bond angles and the amino acid sequence information to predict the secondary structure, and can be further carried forward for the protein folding pattern prediction [36]. Maclin, R., and Shavlik developed a novel hybrid algorithm to improve the performance of the Chou–Fasman algorithm, a method for predicting how globular proteins fold [37]. Reilly et al., in their research work, described a set of artificial copolymers that contains the hydrophobic and hydrophilic units capable of folding into the known 3D structure. Using the recurrent neural networks, they described a model for predicting the possible long-range interactions between regions of the sequence [38]. The work of Battistella and Cechin described the implementation of an ANN method for the prediction of the dihedral angles of protein backbone by considering the physical and chemical properties of the given amino acid sequence alone [39]. A similar approach was also taken by Hamad et al., for the prediction of the 3D structure by using the sequence information along with geometry coordinates as the input layer [40]. Recently, an ANN-based protein structure prediction system AlphaFold was developed by Senior et al., in 2019 [41]. Most recently, the deep learning method has been widely used for solving the protein folding problem. Deep learning methods are important aspects of machine learning algorithms based on ANN with different hierarchical learning levels [42, 43]. Considering the problem of protein folding, deep learning methods such as convolutional neural networks (CNN), recurrent neural networks (RNN), and the basic feed-forward neural network (FFNN) are frequently used [44, 45].

9.8 LIMITATIONS OF CURRENT PROTEIN FOLDING PREDICTION ALGORITHMS

Several approaches have been taken to studying the protein folding process by using recent machine learning methods. The current basis for predicting three-dimensional protein structures by using machine learning tools is the sequence features [46, 47]. Although the ANN-based algorithms have advantages in their performance, there exist specific limitations as given below.

- The folding algorithm does not consider the cellular physiological environment and protein interaction effects that occur in the folding process.

- Accurate prediction of the three-dimensional structure f proteins also varies a lot from the experimental predicted structure.
- There are limitations in the prediction of the native state of the protein structure in the shortest possible time as it requires a lot of computational power in terms of time and space.

9.9 CONCLUSION

Proteins are the cells' real machinery to perform almost all of the functions of the cell by interacting physically in its folded three-dimensional structure. The biotechnology industry in many sectors is mostly dependent on the improved catalytic process of the protein for the understanding of biological function. Therefore, the computational prediction of the folding mechanism involved in case of the proteins, open the door to address the functional aspects. But the major challenge is to predict the protein folding process with good accuracy from the given amino acid sequence. Hence, this can be solved computationally by using ANN and allied algorithms. Artificial neural networks are based on the principle of the neuron. The neuron takes inputs through its many dendrites and sends them to the soma and a nonlinear function is then applied to the signals within the soma. It sums the dendritic inputs together and evaluates the result according to a threshold function. Machine learning techniques, such as genetic algorithms and ant colony optimization, also can be effectively implemented along with the ANN which is good at predicting the protein folding process. Recently, a lot of methodological improvement has been made in the field of protein folding study by using artificial neural networks. However, there is a limitation in the accuracy of these methods, so it is desirable to develop improved novel methods with enhanced protein folding predictions in order to address the important molecular biological process.

BIBLIOGRAPHY

1. Dill, K. A., S. Banu Ozkan, M. Scott Shell, and T. R. Weikl. 2008. The protein folding problem. *Annual Review of Biophysics* 37: 289–316. DOI: 10.1146/annurev.biophys.37.092707.153558, PubMed: 18573083.
2. Chan, H. S., and K. A. Dill. 1993. The protein folding problem. *Physics Today* 46, no. 2: 24–32. DOI: 10.1063/1.881371.
3. Anfinsen, C. B. 1973. Principles that govern the folding of protein chains. *Science* 181, no. 4096: 223–30. DOI: 10.1126/science.181.4096.223, PubMed: 4124164.
4. Levinthal, C. 1968. Are there pathways for protein folding? *Journal de chimie physique* 65: 44–5. DOI: 10.1051/jcp/1968650044.
5. Lifson, S., and A. Roig. 1961. On the theory of helix—coil transition in polypeptides. *The Journal of Chemical Physics* 34, no. 6: 1963–74. DOI: 10.1063/1.1731802.
6. Wetlaufer, D. B. 1973. Nucleation, rapid folding, and globular intrachain regions in proteins. *Proceedings of the National Academy of Sciences* 70, no. 3: 697–701. DOI: 10.1073/pnas.70.3.697.
7. Ptitsyn, O. B., and A. A. Rashin. 1975. A model of myoglobin self-organization. *Biophysical Chemistry* 3, no. 1: 1–20. DOI: 10.1016/0301-4622(75)80033-0.

8. Rackovsky, S., and H. A. Scheraga. 1977. Hydrophobicity, hydrophilicity, and the radial and orientational distributions of residues in native proteins. *Proceedings of the National Academy of Sciences* 74, no. 12: 5248–51. DOI: 10.1073/pnas.74.12.5248.

9. Kuwajima, K. 1977. A folding model of α-lactalbumin deduced from the three-state denaturation mechanism. *Journal of Molecular Biology* 114, no. 2: 241–58. DOI: 10.1016/0022-2836(77)90208-X.

10. Wong, K. P., and C. Tanford. 1973. Denaturation of bovine carbonic Anhydrase B by guanidine Hydrochloride A process involving separable sequential conformational transitions. *Journal of Biological Chemistry* 248, no. 24: 8518–23.

11. Ohgushi, M., and A. Wada. 1983. 'Molten-globule state': a compact form of globular proteins with mobile side-chains. *FEBS Letters* 164, no. 1: 21–4. DOI: 10.1016/0014-5793(83)80010-6, PubMed: 6317443.

12. Ptitsyn, O. B. 1995. Molten globule and protein folding. In. *Advances in Protein Chemistry*. Academic Press 47. DOI: 10.1016/S0065-3233(08)60546-X.

13. Kuwajima, K., Y. Hiraoka, M. Ikeguchi, and S. Sugai. 1985. Comparison of the transient folding intermediates in lysozyme and. alpha.-lactalbumin. *Biochemistry* 24, no. 4: 874–81. DOI: 10.1021/bi00325a010.

14. Ikeguchi, M., K. Kuwajima, M. Mitani, and S. Sugai. 1986. Evidence for identity between the equilibrium unfolding intermediate and a transient folding intermediate: a comparative study of the folding reactions of. alpha.-lactalbumin and lysozyme. *Biochemistry* 25, no. 22: 6965–72. DOI: 10.1021/bi00370a034.

15. Ptitsyn, O. B., R. H. Pain, G. V. Semisotnov, E. Zerovnik, and O. I. Razgulyaev. 1990. Evidence for a molten globule state as a general intermediate in protein folding. *FEBS Letters* 262, no. 1: 20–4. DOI: 10.1016/0014-5793(90)80143-7.

16. Kuwajima, K. 1989. The molten globule state as a clue for understanding the folding and cooperativity of globular-protein structure. *Proteins: Structure, Function, & Genetics* 6, no. 2: 87–103. DOI: 10.1002/prot.340060202.

17. Arai, M., and K. Kuwajima. 2000. Role of the molten globule state in protein folding. *Advances in Protein Chemistry* 53: 209–82. DOI: 10.1016/S0065-3233(00)53005-8.

18. Ahmad, B., and R. H. Khan. 2005. Protein folding: from hypothesis driven to data IVIining. *Pakistan Journal of Biological Sciences* 8, no. 3: 487–92.

19. Yon, J. M. 2001. Protein folding: a perspective for biology, medicine and biotechnology. *Brazilian Journal of Medical & Biological Research = Revista Brasileira de Pesquisas Medicas e Biologicas* 34, no. 4: 419–35. DOI: 10.1590/s0100-879x2001000400001, PubMed: 11285453.

20. Neocleous, C., and C. Schizas. 2002, Apr. Artificial neural network learning: a comparative review. In. *Lecture Notes in Computer Science* Hellenic Conference on Artificial Intelligence. Berlin, Heidelberg: Springer: 300–13. DOI: 10.1007/3-540-46014-4_27.

21. Kim, T. H. 2010, Jun. Pattern recognition using artificial neural network: a review. In. *Communications in Computer & Information Science* International Conference on Information Security and Assurance. Berlin, Heidelberg: Springer: 138–48. DOI: 10.1007/978-3-642-13365-7_14.

22. Da Silva, I. N., D. H. Spatti, R. A. Flauzino, L. H. B. Liboni, and S. F. dos Reis Alves, 2017. Artificial neural network architectures and training processes. In *Artificial Neural Networks*: 21–8. Cham: Springer.

23. Foody, G. M., M. B. McCulloch, and W. B. Yates. 1995. The effect of training set size and composition on artificial neural network classification. *International Journal of Remote Sensing* 16, no. 9: 1707–23. DOI: 10.1080/01431169508954507.

24. Bystroff, C., and D. Baker. 1998. Prediction of local structure in proteins using a library of sequence-structure motifs. *Journal of Molecular Biology* 281, no. 3: 565–77. DOI: 10.1006/jmbi.1998.1943.

25. Khalique, G., and T. Richa, 2018. A survey of the structural parameters used for computational prediction of protein folding process. In *Bioinformatics: Sequences, Structures, Phylogeny*: 255–70. Singapore: Springer.
26. Irbäck, A., C. Peterson, F. Potthast, and O. Sommelius. 1997. Local interactions and protein folding: a three-dimensional off-lattice approach. *The Journal of Chemical Physics* 107, no. 1: 273–82. DOI: 10.1063/1.474357.
27. Irbäck, A., and E. Sandelin. 1998. Local interactions and protein folding: a model study on the square and triangular lattices. *The Journal of Chemical Physics* 108, no. 5: 2245–50. DOI: 10.1063/1.475605.
28. Cotta, C. 2003, Jun. Protein Structure Prediction Using Evolutionary Algorithms Hybridized with Backtracking. In. *Lecture Notes in Computer Science*. Berlin, Heidelberg: Springer: 321–8. DOI: 10.1007/3-540-44869-1_41.
29. Hemeida, A. M., S. A. Hassan, A. A. A. Mohamed, S. Alkhalaf, M. M. Mahmoud, T. Senjyu, and A. B. El-Din. 2020. Nature-inspired algorithms for feed-forward neural network classifiers: a survey of one decade of research. *Ain Shams Engineering Journal* 11, no. 3: 659–75. DOI: 10.1016/j.asej.2020.01.007.
30. Unger, R., and J. Moult. 1993. Genetic algorithms for protein folding simulations. *Journal of Molecular Biology* 231, no. 1: 75–81. DOI: 10.1006/jmbi.1993.1258, PubMed: 8496967.
31. Fidanova, S., and I. Lirkov, 2008, Oct. Ant colony system approach for protein folding. In *Computer Science & Information Technology*. IMCSIT 2008. *International Multiconference on* (pp. 887–891) vol. 2008. DOI: 10.1109/IMCSIT. 2008. PubMed: 4747347. IEEE.
32. Liang, F., and W. H. Wong. 2001. Evolutionary Monte Carlo for protein folding simulations. *The Journal of Chemical Physics* 115, no. 7: 3374–80. DOI: 10.1063/1.1387478.
33. Liang, F., and W. H. Wong. 2001. Evolutionary Monte Carlo for protein folding simulations. *The Journal of Chemical Physics* 115, no. 7: 3374–80. DOI: 10.1063/1.1387478.
34. Chen, M., and W. Q. Huang. 2005. A branch and bound algorithm for the protein folding problem in the HP lattice model. *Genomics, Proteomics & Bioinformatics* 3, no. 4: 225–30. DOI: 10.1016/S1672-0229(05)03031-7.
35. Bohr, H., J. Bohr, S. Brunak, R. M. Cotterill, H. Fredholm, B. Lautrup, and S. B. Petersen. 1990. A novel approach to prediction of the 3-dimensional structures of protein backbones by neural networks. *FEBS Letters* 261, no. 1: 43–6. DOI: 10.1016/0014-5793(90)80632-s, PubMed: 19928342.
36. Sacile, R., and C. Ruggiero, 1993, Oct. Protein folding analysis by an artificial neural network approach. In *Proceedings of the 15th Annual International Conference of the IEEE Engineering in Medicine and Biology Societ*: (1535–6). IEEE.
37. Maclin, R., and J. W. Shavlik. 1993. Using Knowledge-Based Neural Networks to Improve Algorithms: Refining the Chou-Fasman Algorithm for Protein Folding. *Machine Learning* 11, no. 2–3: 195–215. DOI: 10.1007/BF00993077.
38. Reilly, R., M. T. Kechadi, Y. Kuznetosov, E. G. Timoshenko, and K. A. Dawson. 1998. Using recurrent neural networks to predict aspects of 3-D structure of folded copolymer sequences. *Nuovo Cimento D* 20, no. 12: 2565–74.
39. Battistella, E., and A. L. Cechin. 2004, Nov. The protein folding problem solved by a fuzzy inference system extracted from an artificial neural network. In. *Lecture Notes in Computer Science* Ibero-American Conference on Artificial Intelligence. Berlin, Heidelberg: Springer: (474–83). DOI: 10.1007/978-3-540-30498-2_47.
40. Hamad, E. M., N. A. Rawashdeh, M. F. Khanfar, E. N. Al-Qasem et al. 2017. Neural network based prediction of 3D protein structure as a function of enzyme family type and amino acid sequences. Gharabli, S.I. *Jordan Journal of Biological Sciences* 10, no. 2.

41. Senior, A. W., R. Evans, J. Jumper, J. Kirkpatrick, L. Sifre, T. Green, C. Qin, A. Žídek, A. W. R. Nelson, A. Bridgland, H. Penedones, S. Petersen, K. Simonyan, S. Crossan, P. Kohli, D. T. Jones, D. Silver, K. Kavukcuoglu, and D. Hassabis. 2019. Protein structure prediction using multiple deep neural networks in the 13th critical assessment of protein structure prediction (CASP13). *Proteins: Structure, Function, & Bioinformatics* 87, no. 12: 1141–8. DOI: 10.1002/prot.25834.

42. Bengio, Y., A. Courville, and P. Vincent. 2013. Representation learning: a review and new perspectives. *IEEE Transactions on Pattern Analysis & Machine Intelligence* 35, no. 8: 1798–828. DOI: 10.1109/TPAMI.2013.50.

43. Schmidhuber, J. 2015. Deep learning in neural networks: an overview. *Neural Networks* 61: 85–117. DOI: 10.1016/j.neunet.2014.09.003.

44. Torrisi, M., G. Pollastri, and Q. Le. 2020. Deep learning methods in protein structure prediction. *Computational & Structural Biotechnology Journal* 18: 1301–10. DOI: 10.1016/j.csbj.2019.12.011.

45. Giulini, M., and R. Potestio. 2019. A deep learning approach to the structural analysis of proteins. *Interface Focus* 9, no. 3. DOI: 10.1098/rsfs.2019.0003, PubMed: 20190003.

46. Hutson, M. 2019. AI protein-folding algorithms solve structures faster than ever. *Nature.* https://www.nature.com/articles/d41586-019-01357-6

47. Noé, F., G. De Fabritiis, and C. Clementi. 2020. Machine learning for protein folding and dynamics. *Current Opinion in Structural Biology* 60: 77–84. DOI: 10.1016/j.sbi.2019.12.005

10 Application of Machine Learning and Molecular Modeling in Drug Discovery and Cheminformatics

Manish Kumar Tripathi, Sushant Kumar Shrivastava, S. Karthikeyan, Dhiraj Sinha, and Abhigyan Nath

CONTENTS

10.1   INTRODUCTION

The drug discovery process of detecting small molecule ligands which act on the desired target to produce a therapeutic effect has been an extensive and tedious procedure. The identified drug also must be evaluated using multiple *in vitro* and *in vivo* methods to determine its binding affinity against the desired target. Other factors like chemical toxicity and off-target interaction with the unrelated proteins as well as their pharmacogenetic effects, where the genetic variations factor influence the drug

DOI: 10.1201/9781003126164-10

response, are to be considered during drug discovery study [1]. Thus, these multifaceted factors and problems are faced by any drug designer in the drug discovery process. Recently the artificial intelligence approach enables us to develop drugs more efficiently *in silico* with a considerable reduction in time and also in a cost-effective manner before their synthesis and evaluation [2].

Machine learning (ML) has its roots in and is a subclass of artificial intelligence and is presently one of the most sought after and rapidly evolving fields in cheminformatics. The primary application of ML in drug designing is to help researchers to identify and exploit the relationship between the chemical structure and its biological activities or structure-activity relationship (SAR). Due to next-generation technologies [3], there is a boom in the number of sequences, many of which can become fruitful putative drug targets. The discovery of putative drug targets and putative drug molecules is the next essential objective for drug discovery pipelines. Previously many successful drug target prediction methods have been developed using a plethora of sequence, structure, and network-based attributes [4–6]

Along with the ML approach for drug screening, the molecular modeling approach has also been widely applied for drug screening purposes in drug discovery. Thus, molecular modeling has become one of the crucial and indispensable stages in the drug discovery pipelines [7, 8]. Bioinformatics tools and concepts facilitate the reveal of the key genes from the genome data and thus help in the identification of possible protein targets for the drug screening and design process. The objective of molecular modeling is to screen/filter out the most promising hits from a large pool of molecules (from chemical databases) based on a specific criterion. Methods, namely high-throughput virtual screening (HTVS), molecular docking, pharmacophore mapping, and molecular simulation approaches, are currently used for the drug discovery process [9]. Virtual screening (VS) can be performed in two significant ways: structure-based VS, which takes into account the 3D structure of the target molecules to filter out hits [1, 3, 10–13], and ligand-based VS, in which the molecular properties of the ligands are used to sieve out the hits. Molecular docking has been the primary method for VS, but the advent of new powerful and efficient machine learning methods like deep learning has increased the power of VS [14]. Thus, virtual screening searches the large dataset and chemical archives to provide putative drugs based on the active site of the desired target. The molecular simulation approach allows the investigation of both the structural and thermodynamics features of the desired target protein at different levels, which helps in the identification of drug binding sites and to elucidate the drug action mechanism [15].

Thus, further in the chapter, we will focus on and emphasize the different methods of artificial intelligence and molecular modeling, which nowadays are widely used for the drug screening and designing process.

10.2   MACHINE LEARNING METHODS IN CHEMINFORMATICS

Machine learning methods consist of those approaches which can learn from the data without depending on explicit programming. The machine learning models can be used against various tasks such as prediction, classification, regression, etc. Supervised and unsupervised methods are the two main branches of machine learning [16]. Supervised methods are those methods which involve the use of labeled data for training and are used for classification/prediction tasks. Unsupervised methods do not use labeled data and are interested in finding the natural groupings in the dataset.

Deep learning neural networks (DLNN) are now also used for the generation of molecules fulfilling specific criteria/features. Prominent among them are the recurrent neural networks (RNNs) [17], variational autoencoders (VAEs) [18], and generative adversarial networks (GANs) [19]. RNNs are used explicitly for modeling sequential data. Two variants of RNNs that are extensively used are the long short-term memory (LSTM) and gated recurrent unit (GRU) [14]. To learn new higher-order representations from the data makes DLNN superior to other methods [20]. The major disadvantage of traditional neural networks (NNs) is the vanishing gradient problem [21].

Autoencoders are a special type of DLNNs which are trained in an unsupervised fashion. The objective of the autoencoder is to produce its input at its output nodes, in between learning the higher-order representations [5, 22, 23].

Another useful and successful approach to generate ML models with high accuracy is by the implementation of ensemble learners. Ensemble learners consist of several base learners; the results of each of all the base learners are combined to reach a final classification outcome [24, 25]. Bagging [26], boosting [27, 28], and stacking [29–31] are the prominent ensemble learning techniques which have been used widely in many machine learning tasks [32–35]. Bagging usually consists of many decision trees as base learners [36, 37]. Each of the base learners is provided with a bootstrapped sample from the full dataset for training. Consequently, each base learner trains on a different region of the training data. In principle, deficiency in some base learners' learning is compensated by the other base learners. In boosting, the base learners are evolved sequentially during each round of iteration, where more emphasis is given to the hard-to-classify examples. Stacking consist of two stages: training of the base learners takes place in the first stage. The second stage involves a meta learner's training using the outputs of the base learners. Recently, stacked models using DLNN, gradient boosting machines, and random forest have been developed for phospholipidosis inducing potential prediction [29]. The ML technique has also been implemented in drug repurposing, whereby existing drugs can be used for new disease indications [13, 17, 38, 39].

10.2.1 MACHINE LEARNING PLATFORMS

Various well-documented machine learning platforms are available, many of which provide a simple and easy graphical user interface (GUI) which can be used by non-programmers for developing the classification or regression learning models. Prominent among the machine learning platforms is WEKA, which is a Java-based machine learning platform containing the various classification, regression, and clustering algorithms. Apart from WEKA a large number of R and Python packages are available which can be used easily to simulate the different learning algorithms such as Rattle [40], H2O [41], SciKit-learn [42], etc.

10.2.2 Representation of Small Molecules

Most of the machine learning algorithms cannot work on raw genomic and proteomic sequences or the chemical structures. They need to be converted into some numerical value using some mathematical formulation before they can be fed into any learning algorithm. An appropriate numerical representation greatly affects (feature depiction) the training of a machine learning algorithm, because different molecular descriptors present the various facets of the dataset. A wide variety of web servers and standalone programs are available nowadays which can convert protein, genomic, and chemical structural data into useful numerical representations. For example, ChemMine server (https://chemminetools.ucr.edu/) [43] provides the option to obtain JoeLib, OpenBabel, and ChemmineR descriptors for any molecule. Similarly, Chemdes (http://www.scbdd.com/chemdes/) [44] even provides a much broader set of descriptors: Chemopy, CDK, RDKit, Pybel, and BlueDesc descriptors, and also molecular fingerprints (>50 types of molecular fingerprints).

Molecular descriptors can be classified mainly into 1D (e.g. atom counts molecular weight), 2D (e.g. topological indices), and 3D descriptors (3D coordinates of the molecule) [30]. Molecular fingerprints can also be used to represent a set of molecules which are described as vectors containing information for the existence or nonexistence of particular substructures in the molecule [45]. Another useful way to represent the molecules is the transfer learning approach by the use of an autoencoder. The higher-order representations which are constructed using the lower-order features in the hidden layers of the autoencoders can also be used to represent molecules in low dimensions [23]. Alternatively generalized low rank models (GLRM) can also be used for transfer learning of representation for molecules in low dimensions [46]. GLRM proceeds by matrix decomposition and unlike principle component analysis can also be used for data having categorical variables for the purpose of dimensionality reduction [47].

10.2.3 Training Set Creation

For the optimal learning of machine learning algorithms, a diverse representative training set is required [48]. The random splitting of a dataset into separate disjoint training and test set is one of the easiest methods which involve the random selection of a defined percentage of samples into the training and test sets. But random splitting cannot guarantee representative and diverse training and testing sets, as it can get biased towards particular regions of the dataset. To mitigate the shortcomings of random splitting, clustering, and uniform-based sampling methods can be used.

The working principle of K-means clustering is a grouping of samples present in the dataset into a predefined "K" number of clusters. The clusters are created in such a way that the within-group similarity is maximized, and the between-group similarity is minimized. The most common similarity criterion is the Euclidean distance.

Initially, K number of cluster centroids are selected randomly, and the distance between every centroid and the samples is calculated. The next step is the updating of the centroids. Based on minimum Euclidean distance to cluster centroids, clustering proceeds until the movement of samples between the clusters ceases, i.e. the attainment of the convergence criterion. In the representative method of training set creation, one member from each cluster is nominated for the training set, while the rest of the samples are kept for the testing set. A learning algorithm must be provided with all groups of diverse samples so that it can learn all the variations present in the dataset and can generalize well on the unseen testing data. The presence of diverse and representative training and testing sets is required for the proper assessment of learning algorithms.

The other useful method for representative training set creation is the implementation of uniform sampling using the Kennard–Stone (KS) algorithm. In the first step of the KS algorithm, the sample which is closest to the mean of the entire dataset is selected for the representative training set. In further steps, the samples which are farthest from the samples already present in the representative training set are sequentially added. The process is continued until the desired percentage of samples is available in the representative training set.

10.2.4 MODEL EVALUATION METHODS

To evaluate the machine learning algorithms, primarily three evaluation methods are in use: fold-based methods, leave one out cross-validation, K-fold cross-validation, and disjoint training and testing set-based validation.

In K-fold cross-validation, a dataset is segregated into K number of folds and during each training iteration, for training the learning algorithm K − 1 folds are used, while the remaining fold is implemented for its evaluation. Continuation of the entire process takes place until all the folds are implemented as a test fold. Leave one out cross-validation is a special case of K-fold cross-validation, where all samples in the dataset are used for training except one, which is implemented for the purpose of model evaluation. The process is repeated until each sample is assigned as a test case once. Generally, leave one out cross-validation is preferred when the size of the dataset is small as it is computationally intensive (as in each iteration a new model is being constructed). K-fold variation of 10 and 5 is more commonly used. If the dataset size is large, then separate training and testing set-based evaluation is the preferred method. The entire dataset is divided into different disjoint training and testing sets for the purpose of training and model assessment, respectively. Generally, an 80/20 or 70/30 split is preferred for the creation of training and testing sets.

10.2.5 MODEL EVALUATION METRICS

A binary classification problem can be defined as developing a discriminatory model to classify a sample into two classes, either positive class or negative class. The following performance evaluation metrics are used for binary classifiers:

Sensitivity: correctly predicted positive samples percentage

$$\text{Sensitivity} = \frac{TP}{(TP + FN)} \times 100 \qquad (10.1)$$

Specificity: correctly predicted negative samples percentage

$$\text{Specificity} = \frac{TN}{(TN + FP)} \times 100 \qquad (10.2)$$

Accuracy: correctly predicted positive and negative samples

$$\text{Accuracy} = \frac{TP + TN}{TP + FP + TN + FN} \times 100 \qquad (10.3)$$

10.2.6 FEATURE REDUCTION

When representing molecules using some mathematical formulation, a huge number of descriptors are calculated for each molecule which results in the overall increase in the dimension of the representing samples [49]. This may hamper the training of learning algorithms in terms of training time and overall predictive accuracy. A feature reduction/selection algorithm can be implemented in such cases to mitigate the effects of higher dimensionality. Prominent among the feature reduction methods are the wrapper and filter methods, the latter being computationally less expensive [50–52].

10.3 MOLECULAR MODELING METHODS IN CHEMINFORMATICS

The drug discovery methods commonly fall under two broad categories, namely, structure-based drug designing (SBDD) and ligand-based drug designing (LBDD). Structure-based drug designing is performed by using the available structures present in the protein data bank (PDB) database. If crystal structure is not available in the database, then modeling methods are used to predict the structure for drug designing [53]. The modeling software predicts/analyzes the binding site of the small molecule present in the receptor pocket, including the electrostatic field, hydrophobic field, and critical residues involved in hydrogen bond interaction with drug molecules. Further, these binding sites are used for searching the small molecule database for the identification of the lead molecule against the desired target [10, 54, 55]. Virtual screening, molecular docking, and biomolecular simulation methods are presently widely used in structure-based drug design.

The ligand-based drug design method doesn't use the target structure to search for small-molecule libraries. This method is based on the information of well-known molecule binding features against the desired target of interest. Further, these molecules are used to model the pharmacophore, which contains the essential structural features derived from the molecule to bind the target. The pharmacophore modeling and QSAR are the widely used methods for ligand-based drug discovery.

10.3.1 VIRTUAL SCREENING

The exponential rise of new drug targets, and the increase of computational resources have led to accelerating the computer-assisted drug design process by using the cheminformatics techniques such as high-throughput docking, homology search, and pharmacophore search in the database for virtual screening. VS is the easiest method to identify and rank potential drug candidates from compound databases at the early drug discovery phase. It is a quick method for screening potential lead molecules, and numerous effective cases have been recognized using this method [10, 12, 56]. Primarily VS can be done by high-throughput screening (HTS) and *in vitro* assays to screen thousands of compounds against the desired target, which can be a time-consuming process. The advancement of new technologies and computational methods of VS enabled the HTS to become less time consuming and significantly increased the initial screens of compound datasets [57]. This method also helps to entail the computation of compounds to predict which binds well in the target site as compared to other compounds in the given library, instead of testing every compound physically. VS methods are divided into two categories, i.e. structure-based virtual screening (SBVS) and ligand-based virtual screening (LBVS), respectively.

The three main features of HTVS are (i) it filters huge compound libraries into the small sets of active compounds, which will be further used for the experimental study; (ii) it may lead to the optimization of identified lead compounds to increase their affinity against the desired target; (iii) provides a scaffold to develop novel compounds by using the structure-activity relationship.

10.3.2 PHARMACOPHORE MODELING

Pharmacophore is an ensemble of the stearic and electronic features required for the molecular recognition of ligand with a specific biological target. This concept was first introduced by Ehrlich in 1909. The pharmacophore model is generated both in a ligand-based manner and by the structure-based manner. In a ligand-based way, it is generated by superimposing or aligning the set of active molecules and mining the shared features important for the biological activity. It is the key computational strategy of drug screening when the macromolecular target structure is unknown [58]. The two basic steps perform the generation of pharmacophore from the set of multiple ligands is: firstly to generate the conformational space of each ligand which is present in the training set to determine its conformational flexibility, and the other step is to align multiple ligands of the training set to identify the essential common feature. Thus, the generated model will be further used extensively for the drug screening, de novo design, and other applications for drug discovery purposes [59]. Software, namely HypoGen, HipHop, DISCO (https://fluxicon.com/disco/), Tripos (http://www.tripossoftware.com/) PHASE [60], and MOE (https://www.chemcomp.com/Products.htm), is widely used for pharmacophore modeling. The key difference in this software is to use the different algorithms to handle the tractability of ligands and to align them to extract the common features. Despite the advancement to generating pharmacophores by various modeling software, several challenges in the ligand-based pharmacophore modeling still exist, such as modeling the ligand flexibility and properly aligning the ligand molecules.

In the structure-based approach for pharmacophore generation, 3D crystal structure is necessary for the receptor or receptor-ligand complex to generate the pharmacophore model. In this, the pharmacophore model generation is centered on the spatial relationship of the corresponding interaction features present, followed by the assembly of features to generate the model. The structure-based pharmacophore modeling is further divided into two subclasses: structure (without ligand)-based and receptor-ligand based [61]. The receptor-ligand method is based on identifying the ligand binding site into the receptor cavity to determine the significant interaction point between them. Nowadays the e-pharmacophore method, a new approach present in the phase module of Schrödinger, is widely used for drug screening purposes [62]. This method exploits grid-based ligand docking with extra precision (XP) scoring function to calculate the protein-ligand interaction. This method is based on the receptor-ligand complex. It has an advantage over both the ligand- and structure-based approaches by generating the optimized pharmacophore model, which further can be used to rapidly screen a diverse set of bioactive compounds as compared to the conventional structure-based methods.

10.3.3 MOLECULAR DOCKING

This is a theoretical simulation method used to envisage the binding mode and affinity of a ligand molecule against the desired target. This method acts as a promising tool in drug discovery and is widely accepted in the scientific community. Molecular docking helps to validate the experimental data and to find the new ligand molecules against the desired target [63]. The docking methodology is based on two key steps, usually called a searching algorithm and an energy scoring function. The search algorithm helps to generate the possible binding mode of ligand molecules in the active site of the receptor while a scoring function is used to identify the binding strength of the ligand with the receptor and is expressed by dock energy [53]. The docking search algorithm can be divided into three types:

1. *Systematics*: in it, all degrees of freedom are explored for the receptor molecule and ligands are placed into the desired active site. Conformational search, hammerhead algorithm, and incremental construction, etc., are examples of this algorithm.
2. *Random or stochastic*: it includes the Monte Carlo and genetic algorithms (GA). These algorithms are used to envisage the best conformation of the ligand into the binding site of a target via a multistep process.
3. *Molecular dynamics simulation*: it is built on the Newtonian equations of motion and predicts the preferred conformation of the ligand molecule into the protein active site in a time-dependent manner.

Some docking software which works based on the above searching algorithm is GOLD and Autodock (GA), Glide (systematic incremental search techniques),

DOCK (shape-based algorithm), FlexX (incremental construction approaches), and LigandFit (Monte Carlo Simulation).

10.3.4 MOLECULAR SIMULATION APPROACH TO DRUG SCREENING

There are mainly two types of methods used for performing a simulation of the dynamics of molecules, namely metropolis Monte Carlo (MC) and molecular dynamics (MD) simulation. MC simulation performs a random walk in configuration space to generate new configurations [64]. It relies on the Metropolis test, which determines whether the new configuration should be accepted or rejected, which in turn is based on statistical mechanics rather than MD.

MD simulation is a computational process built on the principles of classical mechanics used to predict the behavior of the system as a function of time at the microscopic level. It provides detailed information on the fluctuations or conformational changes in proteins and nucleic acid. MD simulation is widely used for investigating the conformational, structural, and dynamic changes of biological molecules and their complexes.

It is a sampling technique which uses Newton's second law of motion to propagate the system with respect to time. If single-particle i, with a mass m at a position ri (which could be an x, y, or z coordinate in a 3D Cartesian space), then the relation between force and acceleration is given by Equation 10.4:

$$F_i = m_i\, a_i = m_i\, d^2 r_i/dt^2 \tag{10.4}$$

Where Fi and ai represent force and acceleration, on the particle i, and t is the time. Forces which act on the system particles are derived from the gradient of the interaction energy U(r) Equation 10.5:

$$F_i = -\nabla U_i\left(r\right) \tag{10.5}$$

Force exerted by external agent causes translational motion of the spherical molecule. The motion and force applied are explicitly inferred by Newtonian equations of motion. The motion and force applied in the particle system are a set of second-order differential equations in time. The functional form of the motion and force applied is a sum of terms equation (Equation 10.6):

$$m_i\, d^2 r_i/dt^2 = -\nabla_i U_i\left(r_1, r_2, r_3, \ldots .r_N\right)\ I = 1 \ldots .N \tag{10.6}$$

Where m represents the mass of the molecule, and ri represents a vector that locates the atoms on a set of coordinate axes. The force field describes atomic interactions as a sum of bonded and non-bonded interactions. The bonded interaction includes the bond length, bond angle, and bond torsion, whereas the non-bonded interaction includes the van der Waals and electrostatic interactions, respectively.

10.4   CONCLUSION AND FUTURE DIRECTIONS

Machine learning approaches can help to mine complex chemical spaces to sieve out useful information and significantly reduce the time and space for the experimental researchers. With the advent of cheap computing power and deep learning systems, the future of ML in drug discovery pipelines is promising and is going to increase. Moreover, the molecular modeling techniques have been successfully embedded in the automated workflow to screen the large libraries against the desired target. The recent development in computational resources and high-performance computing has also played a key role in the *in silico* drug screening of millions of compounds within a limited time frame. Thus, in the near future and presently also both ML and MD simulation studies can facilitate VS and animal toxicity studies in a comparatively short span of time to identify a lead compound against the desired target.

BIBLIOGRAPHY

1. Wyatt PG, Gilbert IH, Read KD, Fairlamb AH (2011) Target validation: linking target and chemical properties to desired product profile. *Curr Top Med Chem* 11:1275–1283. https://doi.org/10.2174/156802611795429185
2. Vamathevan J, Clark D, Czodrowski P, et al. (2019) Applications of machine learning in drug discovery and development. *Nat Rev Drug Discov* 18:463–477. https://doi.org/10.1038/s41573-019-0024-5
3. Wang JT, Liu W, Tang H, Xie H (2014) Screening drug target proteins based on sequence information. *J Biomed Inform* 49:269–274. https://doi.org/https://doi.org/10.1016/j.jbi.2014.03.009
4. Kumari P, Nath A, Chaube R (2015) Identification of human drug targets using machine-learning algorithms. *Comput Biol Med* 56:175–181. https://doi.org/https://doi.org/10.1016/j.compbiomed.2014.11.008
5. Nath A, Kumari P, Chaube R (2018) Prediction of human drug targets and their interactions using machine learning methods: current and future perspectives. *Methods Mol Biol* 1762:21–30. https://doi.org/10.1007/978-1-4939-7756-7_2
6. Bakheet TM, Doig AJ (2009) Properties and identification of human protein drug targets. *Bioinformatics* 25:451–457. https://doi.org/10.1093/bioinformatics/btp002
7. Sharma P, Tripathi MK, Shrivastava SK (2020) *Cholinesterase as a Target for Drug Development in Alzheimer's Disease BT - Targeting Enzymes for Pharmaceutical Development: Methods and Protocols.* In: Labrou NE (ed). Springer US, New York, NY, pp 257–286.
8. Tripathi MK, Sinha J, Srivastava SK, Kumar D (2019) *Bioinformatics in Skin Cancer: A System Biology Approach to Understanding the Molecular Mechanisms and It's Regulations BT - Skin Aging & Cancer: Ambient UV-R Exposure.* In: Dwivedi A, Agarwal N, Ray L, Tripathi AK (eds). Springer Singapore, Singapore, pp 101–111.
9. Tripathi MK, Sharma P, Tripathi A, et al. (2020) Computational exploration and experimental validation to identify a dual inhibitor of cholinesterase and amyloid-beta for the treatment of Alzheimer's disease. *J Comput Aided Mol Des.* https://doi.org/10.1007/s10822-020-00318-w
10. Xia X (2017) Bioinformatics and drug discovery. *Curr Top Med Chem* 17:1709–1726. https://doi.org/10.2174/1568026617666161116143440
11. Sliwoski G, Kothiwale S, Meiler J, Lowe Jr EW (2013) Computational methods in drug discovery. *Pharmacol Rev* 66:334–395. https://doi.org/10.1124/pr.112.007336

12. Good A (2007) 4.19 - Virtual screening. In: Taylor JB, Triggle DJ (eds) *Comprehensive Medicinal Chemistry II.* Elsevier, Oxford, pp 459–494

13. Napolitano F, Zhao Y, Moreira VM, et al. (2013) Drug repositioning: a machine-learning approach through data integration. *J Cheminform* 5:30. https://doi.org/10.1186/1758 -2946-5-30

14. Carpenter KA, Cohen DS, Jarrell JT, Huang X (2018) Deep learning and virtual drug screening. *Future Med Chem* 10:2557–2567. https://doi.org/10.4155/fmc-2018-0314

15. Tripathi MK, Yasir M, Shrivastava* PS and R (2020) A comparative study to explore the effect of different compounds in immune proteins of human beings against tuberculosis: an in-silico approach. *Curr. Bioinform.* 15:155–164.

16. Larose DT (2006) *Data Mining Methods & Models.* Wiley-Interscience.

17. DiPietro R, Hager GD (2020) Chapter 21 - Deep learning: RNNs and LSTM. In: Zhou SK, Rueckert D, Fichtinger G (eds) *Handbook of Medical Image Computing and Computer Assisted Intervention.* Academic Press, pp 503–519.

18. Simidjievski N, Bodnar C, Tariq I, et al. (2019) Variational autoencoders for cancer data integration: design principles and computational practice. *Front Genet* 10: https://doi.org/10.3389/fgene.2019.01205

19. Kazeminia S, Baur C, Kuijper A, et al. (2020) GANs for medical image analysis. *Artif Intell Med* 101938. https://doi.org/https://doi.org/10.1016/j.artmed.2020.101938

20. Di Gangi M, Lo Bosco G, Rizzo R (2018) Deep learning architectures for prediction of nucleosome positioning from sequences data. *BMC Bioinformatics* 19:418. https://doi.org/10.1186/s12859-018-2386-9

21. Hochreiter S (1998) The vanishing gradient problem during learning recurrent neural nets and problem solutions. *Int J Uncertainty, Fuzziness Knowledge-Based Syst* 6:107–116. https://doi.org/10.1142/S0218488598000094

22. Lopez Pinaya WH, Vieira S, Garcia-Dias R, Mechelli A (2020) Chapter 11 - Autoencoders. In: Mechelli A, Vieira S (eds) *Machine Learning.* Academic Press, pp 193–208

23. Yadav A, Sahu R, Nath A (2020) A representation transfer learning approach for enhanced prediction of growth hormone binding proteins. *Comput Biol Chem* 87:107274. https://doi.org/https://doi.org/10.1016/j.compbiolchem.2020.107274

24. Polikar R (2006) Polikar, R.: Ensemble based systems in decision making. *IEEE Circuit Syst. Mag.* 6, 21–45. Circuits Syst Mag IEEE 6:21–45. https://doi.org/10.1109/MCAS.2006.1688199

25. Polikar R (2009) Ensemble machine learning. *Scholarpedia* 4:2776. https://doi.org/10.4 249/scholarpedia.2776

26. Breiman L (1996) Bagging predictors. *Mach Learn* 24:123–140. https://doi.org/10.1023 /A:1018054314350

27. Schapire R (2002) The boosting approach to machine learning: an overview. Nonlin *Estim Classif Lect Notes* Stat 171. https://doi.org/10.1007/978-0-387-21579-2_9

28. Nath A, Subbiah K (2015) Maximizing lipocalin prediction through balanced and diversified training set and decision fusion. *Comput Biol Chem* 59:101–110. https://doi.org/https://doi.org/10.1016/j.compbiolchem.2015.09.011

29. Nath A, Sahu GK (2019) Exploiting ensemble learning to improve prediction of phospholipidosis inducing potential. *J Theor Biol* 479:37–47. https://doi.org/https://doi.org/10.1016/j.jtbi.2019.07.009

30. Chandrasekaran B, Abed SN, Al-Attraqchi O, et al. (2018) Chapter 21 - Computer-Aided Prediction of Pharmacokinetic (ADMET) Properties. In: Tekade RK (ed) *Dosage Form Design Parameters.* Academic Press, pp 731–755

31. Džeroski S, Ženko B (2004) Is combining classifiers with stacking better than selecting the best one? *Mach Learn* 54:255–273. https://doi.org/10.1023/B:MACH.0000015881 .36452.6e

32. Dey L, Chakraborty S, Mukhopadhyay A (2020) Machine learning techniques for sequence-based prediction of viral–host interactions between SARS-CoV-2 and human proteins. *Biomed J.* https://doi.org/https://doi.org/10.1016/j.bj.2020.08.003

33. Hu S, Zhang C, Chen P, et al. (2019) Predicting drug-target interactions from drug structure and protein sequence using novel convolutional neural networks. *BMC Bioinformatics* 20:689. https://doi.org/10.1186/s12859-019-3263-x

34. Dong L, Yuan Y, Cai Y (2006) Using Bagging classifier to predict protein domain structural class. *J Biomol Struct Dyn* 24:239–242

35. Nath A (2016) Insights into the sequence parameters for halophilic adaptation. *Amino Acids* 48:751–762. https://doi.org/10.1007/s00726-015-2123-x

36. Podgorelec V, Kokol P, Stiglic B, Rozman I (2002) decision trees: an overview and their use in medicine. *J Med Syst* 26:445–463. https://doi.org/10.1023/a:1016409317640

37. Quinlan JR (1986) Induction of decision trees. *Mach Learn* 1:81–106. https://doi.org/10.1023/a:1022643204877

38. Kim E, Choi AS, Nam H (2019) Drug repositioning of herbal compounds via a machine-learning approach. *BMC Bioinformatics* 20:247. https://doi.org/10.1186/s12859-019-2811-8

39. Zhou Y, Wang F, Tang J, et al. Artificial intelligence in COVID-19 drug repurposing. *Lancet Digit Heal.* https://doi.org/10.1016/S2589-7500(20)30192-8

40. Maindonald JH (2012) Data mining with rattle and R: the art of excavating data for knowledge discovery by Graham Williams. *Int Stat Rev* 80:199–200

41. Cook D (2017) Practical machine learning with H2O : powerful, scalable techniques for deep learning and AI.

42. Pedregosa F, Varoquaux G, Gramfort A, et al. (2011) Scikit-learn: machine learning in python. *J Mach Learn Res* 12:2825–2830

43. Backman TWH, Cao Y, Girke T (2011) ChemMine tools: an online service for analyzing and clustering small molecules. *Nucleic Acids Res* 39:W486–W491. https://doi.org/10.1093/nar/gkr320

44. Dong J, Cao D-S, Miao H-Y, et al. (2015) ChemDes: an integrated web-based platform for molecular descriptor and fingerprint computation. *J Cheminform* 7:60. https://doi.org/10.1186/s13321-015-0109-z

45. Cereto-Massagué A, Ojeda MJ, Valls C, et al. (2015) Molecular fingerprint similarity search in virtual screening. *Methods* 71:58–63. https://doi.org/https://doi.org/10.1016/j.ymeth.2014.08.005

46. Schuler A, Liu V, Wan J, et al. (2016) DISCOVERING PATIENT PHENOTYPES USING GENERALIZED LOW RANK MODELS. *Pac Symp Biocomput* 21:144–155

47. Udell M, Horn C, Zadeh R, Boyd S (2016) generalized low rank models. found trends® *Mach Learn* 9:1–118. https://doi.org/10.1561/2200000055

48. Nath A, Subbiah K (2018) The role of pertinently diversified and balanced training as well as testing data sets in achieving the true performance of classifiers in predicting the antifreeze proteins. *Neurocomputing* 272:294–305. https://doi.org/https://doi.org/10.1016/j.neucom.2017.07.004

49. Saeys Y, Inza I, Larrañaga P (2007) A review of feature selection techniques in bioinformatics. *Bioinformatics* 23:2507–2517. https://doi.org/10.1093/bioinformatics/btm344

50. Massart DL, Vandeginste BGM, Deming SN, et al. (2003) Chapter 20 - The multivariate approach. In: *Data Handling in Science and Technology.* Elsevier, pp 319–338.

51. Theodoridis S, Koutroumbas K (2009) Chapter 5 - feature selection. In: Theodoridis S, Koutroumbas K (eds) *Pattern Recognition* (Fourth Edition). Academic Press, Boston, pp 261–322.

52. Hira ZM, Gillies DF (2015) A review of feature selection and feature extraction methods applied on microarray data. *Adv Bioinformatics* 2015:198363. https://doi.org/10.1155/2015/198363

53. Yu W, MacKerell Jr AD (2017) Computer-aided drug design methods. *Methods Mol Biol* 1520:85–106. https://doi.org/10.1007/978-1-4939-6634-9_5

54. Aparoy P, Reddy KK, Reddanna P (2012) Structure and ligand based drug design strategies in the development of novel 5- LOX inhibitors. *Curr Med Chem* 19:3763–3778. https://doi.org/10.2174/092986712801661112

55. Cherian S, Agrawal M, Basu A, et al. (2020) Perspectives for repurposing drugs for the coronavirus disease 2019. *Indian J Med Res* 151:160–171. https://doi.org/10.4103/ijmr.IJMR_585_20

56. Hughes JP, Rees S, Kalindjian SB, Philpott KL (2011) Principles of early drug discovery. *Br J Pharmacol* 162:1239–1249. https://doi.org/10.1111/j.1476-5381.2010.01127.x

57. Lionta E, Spyrou G, Vassilatis DK, Cournia Z (2014) Structure-based virtual screening for drug discovery: principles, applications and recent advances. *Curr Top Med Chem* 14:1923–1938. https://doi.org/10.2174/1568026614666140929124445

58. Daisy P, Singh SK, Vijayalakshmi P, et al. (2011) A database for the predicted pharmacophoric features of medicinal compounds. *Bioinformation* 6:167–168. https://doi.org/10.6026/97320630006167

59. Dror O, Schneidman-Duhovny D, Inbar Y, et al. (2009) Novel approach for efficient pharmacophore-based virtual screening: method and applications. *J Chem Inf Model* 49:2333–2343. https://doi.org/10.1021/ci900263d

60. Dixon SL, Smondyrev AM, Knoll EH, et al. (2006) PHASE: a new engine for pharmacophore perception, 3D QSAR model development, and 3D database screening: 1. Methodology and preliminary results. *J Comput Aided Mol Des* 20:647–671. https://doi.org/10.1007/s10822-006-9087-6

61. Mortier J, Dhakal P, Volkamer A (2018) Truly target-focused pharmacophore modeling: a novel tool for mapping intermolecular surfaces. *Molecules* 23:1959. https://doi.org/10.3390/molecules23081959

62. Salam NK, Nuti R, Sherman W (2009) Novel method for generating structure-based pharmacophores using energetic analysis. *J Chem Inf Model* 49:2356–2368. https://doi.org/10.1021/ci900212v

63. Meng X-Y, Zhang H-X, Mezei M, Cui M (2011) Molecular docking: a powerful approach for structure-based drug discovery. *Curr Comput Aided Drug Des* 7:146–157. https://doi.org/10.2174/157340911795677602

64. Salmaso V, Moro S (2018) Bridging molecular docking to molecular dynamics in exploring ligand-protein recognition process: an overview. *Front. Pharmacol.* 9:923.

11 Role of Advanced Artificial Intelligence Techniques in Bioinformatics

Harshit Bhardwaj, Pradeep Tomar,
Aditi Sakalle, and Uttam Sharma

CONTENTS

11.1   INTRODUCTION

In bioinformatics and artificial intelligence (AI), there has been substantial development in recent decades, and there are more opportunities to consider biological data and challenges. Bioinformatics is an interdisciplinary research area and uses mathematical models, statistical methods and algorithms, and computing power to solve various biological problems [1]. AI is the software system's ability to execute multiple functions related to intelligent agents and simulate human intelligence processes [2]. The use of artificial intelligence techniques in bioinformatics is becoming

DOI: 10.1201/9781003126164-11

increasingly important. Because of the intractability of existing methods or the lack of educated and insightful ways to manipulate biological data, many bioinformatics problems require a new direction to be solved. For example, new methods for extracting gene and protein networks from rapidly proliferating gene expression and proteomic datasets need urgently to be identified.

There is currently very little understanding of how traditional strategies such as clustering, association recognition, and self-organization map analysis of gene expression and proteomic data can lead directly to the reverse engineering of gene networks or metabolic routes. In the first study of traditional enormous datasets (thousands of genes are calculated on a single microscope), such approaches also allow one to classify sufficient and appropriate causal associations between two or more genes or proteins by human experts. Biological information must be paired with simulation methods to derive specific genes from thousands of genomes. For example, it may be conceivable to foresee how a protein flips out of the beginning with any protein sequence algorithms of 20 or so amino acids. But once the sequences become biologically realistic (200 or 300 or more amino acids), existing algorithms for protein folding are effortlessly enticing.

AI is a computer science field that addresses issues that are deemed intractable to computer scientists in heuristic and probabilistic methods. AI approaches are excellent when dealing with problems where the "absolutely correct or best" answer is not needed (a "solid" constraint), but rather a response that is greater than one known currently or reasonable within some specified limits (a "small" limitation). AI approaches are excellent when dealing with problems. Provided that bioinformatics' specific problems have no significant limits, many bioinformatics problems can be addressed through AI techniques.

AI is a machine's capacity to interpret, create, and carry out activities through its training [3]. Numerous biological questions have been addressed with bioinformatics methods [4]. The Structural Bioinformatics Tools apply AI [5-11] and have efficient methodologies for the development, through silicon-based instruments, of active new compounds to prevent cancer [12-17]. Bioinformatics methods are used for evaluating biological data for rational reasons. Entire genome sequencing programs generate large volumes of biological data, and bioinformatics methods are useful for resolving and describing the data.

Many biological problems have been solved using AIs and bioinformatics techniques for efficient gene prediction algorithms, electronic drug creation, interactional protein-protein tests, comprehensive association genome tests, and future generation software development sequences. AI provides a broad scope of bioinformatics applications [18, 19]. By altering its code according to the input data, AI performance improves. The computer carries out the activities as humans do when implementing AI, and can enhance human reasoning and emotions [20-22]. AI is known as generalized AI and applied AI. Heuristic approaches and AI demonstrate the opportunity to transform biotechnology, pharmacology, and medicine into bioinformatics' latest and future applications [23, 24]. Apart from patients' diagnostics, applied AI now helps forecast custom drugs by examining their genomic data. In contrast to algorithms without any AI experience, the AI base has the potential to interpret data effectively.

AI algorithms, including linear regression, logistic regression, and a bioinformatics K-nearest neighbor algorithm, increase computational biological performance to overcome biological problems [25, 26].

11.2 BIOINFORMATICS: ANALYZING LIFE DATA AT THE MOLECULAR LEVEL

Living cells consist of a wide range of distinct molecules of multiple functionalities. Apart from water, biopolymers, i.e., molecules containing long-chain elementary components bound by covalent connections, are the primary constituents that can be viewed as text on fixed alphabets at an abstract level. Four major biopolymer groups exist: DNA [27, 28], RNA [29, 30], proteins [31, 32], and glycans [33, 34].

11.2.1 DNA

The most common form of macromolecule is a genetically modifiable DNA molecule. DNA consists of four components (bases), consisting of four letters (A, T, G, and C), where most of them align to form a double strand (A, T, and C, G are the canonical matches) [35]. The DNA test is highly organized, covering regions that encode genes and others that express these genes for control. In general, DNA is similar in all its cells, and it can be altered with so-called epigenetic factors such as DNA methylation or histone shifts [36]. It is also well known that DNA can produce genetically modified mutations (polymorphism) between humans. One of the big DNA problems is annotation, which explores the functional information encoded by the nucleotide sequence: genes and other genomic units. For essential organisms, laboratories regularly use offshoring information pipelines, but it remains a complicated challenge [37, 38].

11.2.2 RNA

The second macromolecule, RNA, also consists of 4 (A, U, G, and C) nucleotides researched in the last 15 years and shows a central role in the regulatory process [39]. RNA is known to be a source of earthly life. The primary virus types consist of RNA, typically a single strand folded on itself, forming different space structures related to its purpose. The separation is a big obstacle in RNA research expressed in laboratory settings, development stages, or pathologic and unsafe disorders. Beyond the expression of individual genes, specialized researchers investigate gene sets or genetic networks either as co-expression networks or as biological pathways with distinct signals [40].

The lower costs for these two nucleic acid groups for sequence technology chains profoundly influence biological science culture. In turn, sequencing programs have typically been small enterprises and are now being undertaken by each lab.

11.2.3 PROTEINS

Proteins, the third group of macromolecules, are the major players in the cell whose roles vary from cell formation (e.g., collagen) to biochemical catalysis (enzymes), transportation, cell replication, and DNA [41]. Protein controls the continuous life dynamics component: essentially, all biochemical reactions depend on protein expression at continuous rates. We know it's essential to understand the structure of proteins and their role. Compared to nuclear polymers, even the protein sequence is difficult to attain because proteins are mostly processed by post-translation (for example, glycosylation of more than 70% on eukaryotic proteins). The sequence of proteins is difficult to obtain [42]. Only appropriate studies will generate protein structures, and this structure estimate from a sequence is a long-term demanding task in bioinformatics. The production of new pharmaceutical products is another especially active area of research due to its significance in human health.

11.2.4 GLYCANS

Polysaccharides consist of long chains of monosaccharide atoms, usually coupled with a covalent glycoside bond, which is the final biopolymer form, sometimes referred to as single sugars (mostly four forms, but hundreds of structural isomers and versions have been identified) [43]. Polysaccharide also provides an energy source for the body (for example, starch can become glucose) or may play a structural role. Proteins (glycoprotein or proteoglycans) and lipids (glycolipids) are predominantly linked to glycans. After glycan conversion, the modification of certain protein amino acids is known as glycosylation. Glycan structures (sequencing), including the recognition of glycosylated sites in the protein, are still a complicated experimental procedure. In comparison to other biopolymers, glycobiological developments are recent [44].

11.3 APPLICATION OF AI IN BIOINFORMATICS

For at least two reasons, artificial intelligence is closely involved in the advancement of bioinformatics. The use of neural networks [45], evolutionary methods including genomic and growth models [46], immunological computing [47], and various modeling approaches like the optimization of swarms or ants and logics is done for solving problems such as the fugitive method or the principle of possibilities. In the second date to the start of the century biology became a data-intense field of science and various instruments were developed, which scanned life on a molecular level. Currently, biology is not just a data science but rather an information science. Artificial intelligence is a great technology area for research on the representation of information, logic, and machine learning. Also, bioinformatics addresses various NP-hard topics that are of concern to artificial intelligence.

In general, the four significant areas of study in bioinformatics concern statistics, algorithms, modeling, and optimization problems. The first examines challenges resulting from the collection of chaotic statistical data or the globalization of populations with vector elements such as population genetics. Although artificial intelligence is interested in these issues, the inputs in this area originate from existing statistical methods. The second direction resulted in several innovations, as sequences are the primary database. The same expression data are given now in the form of sequences that quantify such compounds' quantity. In this domain, it is essential to find linear analytical algorithms because of the scale of data – often several gigabytes. In achieving this goal, indexation strategies play a significant role. The last two tracks deal entirely with artificial intelligence methods, which also raise some fascinating scientific problems.

The complexity of structures and mechanisms is distinctly visible from physics to chemistry and then biology. Living organisms first hold knowledge in complex molecules safely, and they have several symbolic devices (for example, polymerases, robotics, and spliceosome) for the deciphering of the corresponding code. Biology is the study of partnerships and attachments par excellence. It examines a range of artifacts linked to multiple connections at each level in a hierarchical organizational system. The biological records contain a vast volume of data, and the annotation of these data appears to be controversial for researchers. AI systems in the bioinformatics field are capable of capturing data to rational conclusions [48]. The fusion of AI and bioinformatics will effectively forecast various organisms' simulations, biological sequence annotation, estimation of drugs, computer screening, and gene prediction. The significant contribution of AI in bioinformatics studies relies on combining models and knowledge-based structures to solve biological issues.

Advances in AI and bioinformatics are relevant in health sciences and contribute to clinic bioinformatics, high-performance screening, disease prevention, epidemiology, and immunosuppression. The production of new vaccines is increasingly complicated as mutations in microbes and viruses are increasingly prevalent.. Computer systems will scan the target vaccines from >20,000 proteins to >100,000 proteins from influenza per day. The data generated can be analyzed with the aid of various methods and findings. AI's bioinformatics performance has been widely used to solve various biological problems with algorithms and methodologies, including neural networks, probabilistic models, decision trees, virtual automatons, hybrid methods, and genetic algorithms [49, 50].

Provided that several problems are not highly restricting in bioinformatics, a few bioinformatics problems have ample space for applying AI techniques. Some of the general AI techniques are discussed in this paper for the bioinformatics application. Firstly, classical techniques of symbolic machine learning are implemented, including examples of bioinformatics applications in the prediction of secondary protein structures and viral protease cleavability, using nearest neighbor and tree recognition approach. Next, a variety of implementations are defined, and neural networks will be implemented. Thirdly, genetic algorithms with multi-sequence alignment and RNA folding prediction will be used.

In bioinformatics, there are problem areas where AI can play an important role.

11.4 SYMBOLIC MACHINE LEARNING

11.4.1 NEAREST NEIGHBOR APPROACHES IN BIOINFORMATICS

Bioinformatics researchers usually aim to find the bio sequences most comparable to a separate bio sequence. Those bio sequences include hundreds and thousands of "attributes," i.e., positions that help recognize bio sequence similarities. Usually, far more attributes exist than sequences, and therefore it would be entirely unpredictable to select the characteristics to use for research. For this reason, nearest neighbor estimates typically take all details in all locations into consideration before a missed attribute is allocated [51].

11.4.2 APPLICATION IN VIRAL PROTEASE CLEAVAGE PREDICTION

Symbolic machine learning and classification have many uses in bioinformatics. One primary function applies to the estimation of viral protease cleavage. Virions of intact HIV and HCV are endocytic (inserted in a cell) by cell receptors in humans. A single-stranded RNA chain, usually between 8 and 12 kilobases, with at least nine genes containing genes to generate, is used for "retroviruses."

Symbolic machine learning techniques have not been commonly used in bioinformatics. Still, linear programming techniques have also been used to diagnose breast cancer, and more recently (see Section 11.3), gene expression data have begun to be mined using decision-tab approaches. Recently BIOKDD01 (Bioinformatics Data Mining Workshop8) was a welcome creation held in San Francisco in August of 2019. It shows an ever-rising interest in applying machine-learning techniques to historically tricky problems in bioinformatics. There is a growing understanding of the need for more educated biological data analyses, and a range of valuable online guides also exist to help scientists understand specific techniques. Holsheimer and Siebes (1991) offer a clear systematic and historical introduction to the field of data mining. For microarray data analysis clustering, even secondary structure estimation, the closest approaches were used.

11.5 NEURAL NETWORKS IN BIOINFORMATICS

Microarray (gene chip) data are generated over a series of time stages, often when the body is exposed to stimulation, or through a series (one cycle) of people who fall into various groups (either) by calculating gene expression values or "activation" values. This analysis typically analyses thousands of genes (fields, attributes) and just tens of records for the data on time gene expression or tens of nontemporal results. The purpose of modeling and evaluating the data on temporal gene expression is to examine how excitations and inhibitions occur between genes that comprise an organism's gene-regulatory network. By studying the associations between genes from time to time, a gene regulation network may be deduced. Finding an ideal or even approximate gene regulatory mechanism will increase the comprehension of the regulatory system. For example, drugs can be engineered to interact with or facilitate gene expression across the network at some stages.

For non-temporary data, the aim is to distinguish a subset of genes that contribute to a particular class of people (e.g., diseased or normal) from the thousands of genes calculated for a person and use these genes to determine an individual's class using a small test of the potential diagnosis. Indeed, it is not easy to classify gene regulatory variations within microarray data, particularly for nontemporal data. Interestingly, in these fields, there has been no effort to incorporate standard input ANNs. Perhaps the representation of such information was not noticeable to an FFBP network.

Some neural network applications in bioinformatics, particularly their possible novel use in gene expression data modeling, have been identified. Neural networks are adaptive learning platforms and have been used in several different fields and it should also surprise that they have had some bioinformatics success. Their learning properties and the ease of use for new fields mean that they are already being used in many other bioinformatics areas.

Doctors with prospective palm-held computers, can help determine whether patients suffer from disease during their treatment without providing a detailed explanation for their estimation. By using the first estimate on a specific disease by the palm-held computer, depending on the patient's gene expression profile, the ANNs can configure specific palm-holder machines. For more detailed research, referrals can be provided to specialist consultants. Drug companies that produce new medicines to cure disease require comprehensive explanations of why a disease evolves and advances its actions at the disease's gene-level.

The findings reveal that the ANNs in their approach to data processing are "softer" and, in many cases, more reliable in their predictions. On the other hand, symbolic methods create symbolic awareness and explanations. While ANNs have been widely used in protein folding prediction, the challenge is that biologists still want to know whether the protecting drugs are predicted when needed. Yet ANNs have been highly useful and usable only where prediction and initial classification are necessary.

11.6   EVOLUTIONARY COMPUTATION IN BIOINFORMATICS

The concept of using an (artificial) evolutionary method to create solutions for the complex optimization problem is based on evolutionary computation. The benefit of such a method is that suitable (though not necessarily optimal) solutions can be found easily in huge quantities. The population of emerging candidate solutions can also be spread through a few parallel computer processors, which massively increases the time taken to identify suitable solutions to huge problems.

In the field of bioinformatics, biological metaphors for evolutionary algorithms are important. We are attempting to create solutions instead of developing or seeking an answer to our problem. As in natural growth, random shift and selection directed to better performance solutions are the two main ingredients in this process. Therefore, the art of using evolutionary algorithms is to demonstrate how candidate solutions can be used to balance the sum by making random modifications ("mutations" and "cruise"). Genetic algorithms are commonly known to be used for RNA folding. They have also been used to design simulations of networks in biochemistry and DNA.

11.7 DEEP LEARNING IN INFORMATICS

Data have been fundamental to life sciences over the past few decades. Biomedicine and, more precisely, bioinformatics is a prime illustration of this. The implementation of automated networks that are computerized and networked and the remarkable developments in non-invasive data collection technology position data at the forefront of these disciplines. It is anticipated that data would become the gateway to exploring new physiological information in all scales, opening the doors to medical developments unreachable to date. This kind of transition from information data is a normal aim for ML. To be recognized in bioinformatics and biomedicine, ML-based methods must be trusted, especially in medical practice. The task facing ML to achieve that required trustworthiness is to become explainable and [48,49] interpretable. A pressing social concern strengthens this challenge's importance: executing the latest Directive for General Data Security Policy of the European Union (GDPR). It is supposed to, with minimal difference between European countries, be applied in 2018. Its Reference Post referring to automatic human decision-making, including profiling.

The decisions taken by programmed or artificially intelligent algorithmic systems are clarified [50]. In biomedical decision-making, the GDPR directive makes model interpretability a primary consideration. It involves ML directly and should particularly worry those who want to see DL as being used, outside fundamental science, in medical practice. The new existence granted to connectionism in the ML is precisely another explanation for the explainability and interpretability of late, divisive, and intensely contested ML topics. DL methods are, in general, severe cases of black-box models. Being such a success story for ML, the lack of interpretability of DL models has become an urgent concern in the field via recent literature. This is a dilemma with no simple answer, considering the complexity of producing these solutions. Very complicated structures also become transparent somehow. In [51], there is the teaching of adversaries. Recently, a method has been suggested. Model neurons are endowed with human-interpretable concepts that will map findings back to powerful neurons and interpretable representations, offering an overview of how models render their predictions. Recently, interpretability criteria based on the deep-network analysis in the information plane [52] have been identified. It has also been claimed that lack of accountability is one of the critical obstacles to acceptance and the implementation of ML methods in general and DL methods, particularly in medicine [43]. Che, and co-workers express their views; in [53], they recommend gradient enhancing decision trees to derive interpretable information from a deep professional network. In similar work [54], deep models are regularized such that their estimates of class likelihood can be modeled with minimal loss by trees of judgment with few nodes. It is much more tractable and intuitive to interpret these trees from the original models.

11.8 FUTURE TRENDS

The application of DL approaches to biomedical and bioinformatic concerns is a multifaceted technique. There is a problem. in providing a detailed view of potential developments in the country and of the numerous significant challenges. We are confronted in real-world situations with such programs. However, several potential developments have been highlighted in biomedicine [52] and bioinformatics [8]. DNNs are cited for biomedicine to be of possible interest to areas as diverse as semantic linkage, the creation of biomarkers, and the discovery of drugs (structural study and hypothesis). The formulation through the study of DNN abstract learned representations, clinical advice and analysis of transcriptomic results is investigating properly to encode raw for bioinformatics. Instead of human-processed functionality and multi-modal data types and learning relevant characteristics for DNNs, these multi-modal or basic types are a possible obstacle. Please notice that previous remarks do not include medical applications in hospital facilities and clinical conditions. Explainability and interpretability in these areas, and as stated in the previous pages, for DL, are significant obstacles. Despite substantial studies of neural network perception, we are in clinical environments that date back many decades [51] and expand to more recent work [53]. The first threads of techniques capable of translating the intricate inner workings are now woven with profound architecture. This suggests that this is perhaps both a positive forward research pattern and a strong future analysis trend – a far-from-trivial mission. A new analysis by Google Brain researchers illustrates the above [49]. This shows the shocking consequence of the lack of sensitivity of the local model's explanations to the deep parameter values for a neural network (DNN). The report concludes that "the DNN architecture. is right before input and can grab low-level input with random initialization". As stated, the coherent incorporation of multi-modal data [50] is another possible future trend and definite challenge for DL. In one of the experiments in the ESANN 2018 special session covered by this tutorial [41], a case of multi-omics data integration is discussed. It provides a novel super layered model of the neural network called the cross-modal neural network. Surprisingly, it is supposed to do well in situations with a small range of usable training samples. A final glimpse of the future is offered in [42], where a brain dynamics-model-free approach to reinforcement learning to regulate the reaction properties of biological neural networks is presented.

11.9 CONCLUSION

An early, unexplored stage is still the implementation of AI technologies in bioinformatics. However, the paper's examples show the enormous potential of AI techniques for contributing to our expertise in this area usefully and functionally. Free software allows researchers in bioinformatics to use the required resources. While using the free softwares, some expense is possible for studying how these resources can be used, but the benefits will surely be worth it. Although this analysis discusses the most common AI techniques, several other techniques are still covered: hidden Markov models, Bayesian networks and probabilistic reasoning, fuzzy logic, temporary logic, and spatial logic.

Furthermore, bioinformatics gives AI researchers the chance to validate their technologies in a different environment. The discrepancy between AI's past trouble areas and the idea that data are rising unbelievably rapidly with our abilities to interpret and model the data stretching to the maximum. We must be more prepared for an educated and insightful study of the increasing amount of data, and that is what AI, and bioinformaticians, will do for us eventually.

BIBLIOGRAPHY

1. Sehgal, S. A., Hammad, M. A., Tahir, R. A., Akram, H. N., and Ahmad, F.. Current therapeutic molecules and targets in neurodegenerative diseases based on in silico drug design. *Current neuropharmacology* 16, no. 6, 649–663, 2018.
2. Sehgal, S. A., Khattak, N. A., and Mir, A.. Structural, phylogenetic and docking studies of D-amino acid oxidase activator (DAOA), a candidate schizophrenia gene. *Theoretical Biology and Medical Modelling* 10, no. 1, 3, 2013.
3. Sehgal, S. A., Mannan, S., Kanwal, S., Naveed I., and Mir, A.. Adaptive evolution and elucidating the potential inhibitor against schizophrenia to target DAOA (G72) isoforms. *Drug Design, Development and Therapy* 9, 3471, 2015.
4. Sehgal, S. A., Mannan, S., & Ali, S., Pharmacoinformatic and molecular docking studies reveal potential novel antidepressants against neurodegenerative disorders by targeting HSPB8. *Drug Design, Development and Therapy, 10,* 1605, 2016.
5. Sehgal, S. A., Hassan, M., & Rashid, S., Pharmacoinformatics elucidation of potential drug targets against migraine to target ion channel protein KCNK18. *Drug Design, Development and Therapy, 8,* 571, 2014.
6. Sehgal, S. A., Pharmacoinformatics and molecular docking studies reveal potential novel Proline Dehydrogenase (PRODH) compounds for Schizophrenia inhibition. *Medicinal Chemistry Research*, 26(2), 314–326, 2017.
7. Sehgal, S. A. Pharmacoinformatics, adaptive evolution, and elucidation of six novel compounds for schizophrenia treatment by targeting DAOA (G72) isoforms. *BioMed Research International, 2017,* 2017.
8. Tahir, R. A., & Sehgal, S. A., Pharmacoinformatics and molecular docking studies reveal potential novel compounds against schizophrenia by target SYN II. *Combinatorial Chemistry & High Throughput Screening, 21*(3), 175–181, 2018.
9. Jamil, F., Ali, A., & Sehgal, S. A., Comparative modeling, molecular docking, and revealing of potential binding pockets of RASSF2; a candidate cancer gene. *Interdisciplinary Sciences: Computational Life Sciences, 9*(2), 214–223, 2017.
10. Sehgal, S. A., Tahir, R. A., Shafique, S., Hassan, M., & Rashid, S., Molecular modeling and docking analysis of CYP1A1 associated with head and neck cancer to explore its binding regions. *J Theor Comput Sci*, 1(112), 2, 2014.
11. Sehgal, S. A., Kanwal, S., Tahir, R. A., Khalid, Z., & Hammad, M. A., In silico elucidation of potential drug target sites of the Thumb Index Fold Protein, Wnt-8b. *Tropical Journal of Pharmaceutical Research, 17*(3), 491–497, 2018.
12. Tahir, R. A., Sehgal, S. A., Khattak, N. A., Khattak, J. Z. K., & Mir, A., Tumor necrosis factor receptor superfamily 10B (TNFRSF10B): an insight from structure modeling to virtual screening for designing drug against head and neck cancer. *Theoretical Biology and Medical Modelling, 10*(1), 38, 2013.
13. Dali, Y., Abbasi, S. M., Khan, S. A. F., Larra, S. A., Rasool, R., Ain, Q. T., & Jafar, T. H., Computational drug design and exploration of potent phytochemicals against cancer through in silico approaches. *Biomedical Letters*, 5(1), 21–26, 2019.
14. Xia, F., Shukla, M., Brettin, T., Garcia-Cardona, C., Cohn, J., Allen, J. E., ... & Stahlberg, E. A., Predicting tumor cell line response to drug pairs with deep learning. *BMC Bioinformatics, 19*(18), 71–79, 2018.

15. Sun, D., Wang, M., & Li, A., A multimodal deep neural network for human breast cancer prognosis prediction by integrating multi-dimensional data. *IEEE/ACM Transactions on Computational Biology and Bioinformatics*, *16*(3), 841–850, 2018.

16. Maréchal, E., Chemogenomics: a discipline at the crossroad of high throughput technologies, biomarker research, combinatorial chemistry, genomics, cheminformatics, bioinformatics and artificial intelligence. *Combinatorial Chemistry & High Throughput Screening*, *11*(8), 583–586, 2008.

17. Coudray, N., Ocampo, P. S., Sakellaropoulos, T., Narula, N., Snuderl, M., Fenyö, D., & Tsirigos, A., Classification and mutation prediction from non–small cell lung cancer histopathology images using deep learning. *Nature Medicine*, *24*(10), 1559–1567, 2018.

18. Cannata, N., Schröder, M., Marangoni, R., & Romano, P., A Semantic Web for bioinformatics: goals, tools, systems, applications, 2008.

19. Yu, K. H., Beam, A. L., & Kohane, I. S., Artificial intelligence in healthcare. *Nature Biomedical Engineering*, *2*(10), 719–731, 2018.

20. Keedwell, E., & Narayanan, A., *Intelligent Bioinformatics: The Application of Artificial Intelligence Techniques to Bioinformatics Problems*. John Wiley & Sons, 2005.

21. Ghahramani, Z., Probabilistic machine learning and artificial intelligence. *Nature*, *521*(7553), 452–459, 2015.

22. Eden, A. H., Steinhart, E., Pearce, D., & Moor, J. H., Singularity hypotheses: an overview. In *Singularity Hypotheses* (pp. 1–12). Springer, Berlin, Heidelberg, 2012.

23. Saeys, Y., Inza, I., & Larrañaga, P., A review of feature selection techniques in bioinformatics. *Bioinformatics*, *23*(19), 2507–2517, 2007.

24. Larranaga, P., Calvo, B., Santana, R., Bielza, C., Galdiano, J., Inza, I., & Robles, V., Machine learning in bioinformatics. *Briefings in Bioinformatics*, *7*(1), 86–112, 2006.

25. Castiglione, F., Pappalardo, F., Bernaschi, M., & Motta, S., Optimization of HAART with genetic algorithms and agent-based models of HIV infection. *Bioinformatics*, *23*(24), 3350–3355, 2007.

26. Watson, J. D., Laskowski, R. A., & Thornton, J. M., Predicting protein function from sequence and structural data. *Current Opinion in Structural Biology*, *15*(3), 275–284, 2005.

27. Escobar-Zepeda, A., Vera-Ponce de Leon, A., & Sanchez-Flores, A. The road to metagenomics: from microbiology to DNA sequencing technologies and bioinformatics. *Frontiers in Genetics* 6, 348, 2015

28. Bianchi, L., & Liò, P.. Forensic DNA and bioinformatics. *Briefings in Bioinformatics* 8(2), 117–128, 2007.

29. Marz, M., Beerenwinkel, N., Drosten, C., Fricke, M., Frishman, D., Hofacker, I. L., Hoffmann, D. et al. Challenges in RNA virus bioinformatics. *Bioinformatics* 30(13), 1793–1799, 2014.

30. Huang, X., S. Liu, L. Wu, M. Jiang, & Y. Hou. High throughput single cell RNA sequencing, bioinformatics analysis and applications. In *Single Cell Biomedicine*, pp. 33–43. Springer, Singapore, 2018.

31. Park, S. K., Bae, D. H., & Rhee, K. C.. Soy protein biopolymers cross-linked with glutaraldehyde. *Journal of the American Oil Chemists' Society* 77(8), 879–884, 2000.

32. Cieplak, P., Cornell, W. D., Bayly, C., & Kollman, P. A.. Application of the multimolecule and multiconformational RESP methodology to biopolymers: charge derivation for DNA, RNA, and proteins. *Journal of Computational Chemistry* 16(11), 1357–1377, 1995.

33. Bertozzi, C. R., & Rabuka, D.. *Structural basis of glycan diversity*, 2010.

34. Geissner, A., & Seeberger, P. H.. Glycan arrays: from basic biochemical research to bioanalytical and biomedical applications. *Annual Review of Analytical Chemistry* 9, 223–247, 2016.

35. Machanick, P., & Bailey, T. L.. MEME-ChIP: motif analysis of large DNA datasets. *Bioinformatics* 27(12), 1696–1697, 2011.

36. Hasan, M. M., Manavalan, B., Khatun, M. S., & Kurata, H. i4mC-ROSE, a bioinformatics tool for the identification of DNA N4-methylcytosine sites in the Rosaceae genome. *International Journal of Biological Macromolecules* 157, 752–758, 2020.

37. Kress, W. J., & Erickson, D. L.. DNA barcodes: genes, genomics, and bioinformatics. *Proceedings of the National Academy of Sciences* 105(8), 2761–2762, 2008.

38. Masoudi-Nejad, A., Tonomura, K., Kawashima, S., Moriya, Y., Suzuki, M., Itoh, M., Kanehisa, M., Endo, T., & Goto, S.. EGassembler: online bioinformatics service for large-scale processing, clustering and assembling ESTs and genomic DNA fragments. *Nucleic Acids Research* 34, no. suppl_2, W459–W462, 2006.

39. Nikogosyan, D. N., & Gurzadyan, G. G.. Two-quantum photoprocesses in DNA and RNA biopolymers under powerful picosecond laser UV irradiation. *Laser Chemistry* 4(1–6), 297–303, 1984.

40. Draper, D. E. A guide to ions and RNA structure. *Rna* 10(3), 335–343, 2004.

41. Peppas, N. A. Devices based on intelligent biopolymers for oral protein delivery. *International Journal of Pharmaceutics* 277(1–2), 11–17, 2004.

42. Matveev, Yu I., Ya Grinberg, V., & Tolstoguzov, V. B.. The plasticizing effect of water on proteins, polysaccharides and their mixtures. Glassy state of biopolymers, food and seeds. *Food Hydrocolloids* 14(5), 425–437, 2000.

43. Armstrong, Z., & Withers, S. G. Synthesis of glycans and glycopolymers through engineered enzymes. *Biopolymers* 99(10), 666–674, 2013.

44. Lowe, J. B., & Marth, J. D.. A genetic approach to mammalian glycan function. *Annual Review of Biochemistry* 72(1), 643–691, 2003.

45. Zeng, H., Edwards, M. D., Liu, G., & Gifford, D. K.. Convolutional neural network architectures for predicting DNA–protein binding. *Bioinformatics* 32(12), i121–i127, 2016.

46. Posada, D., Crandall, K. A., & Holmes, E. C.. Recombination in evolutionary genomics. *Annual Review of Genetics* 36(1), 75–97, 2002.

47. Rapin, N., Lund, O., Bernaschi, M., & Castiglione, F.. Computational immunology meets bioinformatics: the use of prediction tools for molecular binding in the simulation of the immune system. *PLoS One* 5(4), e9862, 2010.

48. Sajda, P., Machine learning for detection and diagnosis of disease. *Annu. Rev. Biomed. Eng.*, *8*, 537–565, 2006.

49. Tahir, R. A., Wu, H., Rizwan, M. A., Jafar, T. H., Saleem, S., & Sehgal, S. A., Immunoinformatics and molecular docking studies reveal potential epitope-based peptide vaccine against DENV-NS3 protein. *Journal of Theoretical Biology*, *459*, 162–170, 2018.

50. Lee, J. W., Lee, J. B., Park, M., & Song, S. H., An extensive comparison of recent classification tools applied to microarray data. *Computational Statistics & Data Analysis*, *48*(4), 869–885, 2005.

51. Zhang, M-L., and Zhou Z-H.. A k-nearest neighbor based algorithm for multi-label classification. In *2005 IEEE International Conference on Granular Computing*, vol. 2, pp. 718–721. IEEE, 2005.

52. Altman, R. B., Challenges for intelligent systems in biology. *IEEE Intelligent Systems*, *16*(6), 14–18, 2001.

53. Narayanan, A., Wu, X., & Yang, Z. R., Mining viral protease data to extract cleavage knowledge. *Bioinformatics*, *18*(suppl_1), S5–S13, 2002.

12 A Bioinformatics Perspective on Artificial Intelligence in Healthcare and Diagnosis
Applications, Implications, and Limitations

Anuradha Bhardwaj, Arun Solanki, and Vikrant Nain

CONTENTS

DOI: 10.1201/9781003126164-12

12.1 INTRODUCTION

Data collection has now become an essential part of many organizations of different domains. The good quality and quantity of data serve as the basis of new developments. It also helps to better organize and fine-tune work strategies for optimization aiming for better future outcomes. Bioinformatics in healthcare is a unique combination of this data supported with health-related information to produce knowledge. Current intelligent algorithms and learning tools allow us to exploit the available big data to make future predictions. This potential of the vast catalogs of data motivates us to produce, collect, and analyze more data. In this era of data where our digital universe is expanding every second, the amount of data a human produce per day through various activities like social activities, sports, the internet, etc., is truly mind-boggling. Our technological advancements help us to record and analyze such data. As the growth rate of this digital data is very high, it is a matter of concern for us to store, manage, and analyze this big data.

The multi-dimensional healthcare sector is a critical pillar of a nation. It primarily focuses on prevention, diagnosis, and treatments. With the advent of computer systems and electronic records, data digitalization, high-throughput techniques, modern instruments, toolkits, etc., the healthcare sector has become more data-intensive and flooded with tons of data. In the year 2003, the term electronic health records (EHR) was coined by the National Academies of Sciences, Engineering, and Medicine to record patients' standard physical and mental health conditions to be stored as digital data for future reference (Reisman, 2017). Due to a wide gap between the number of healthcare professionals and patients, our global health industry is struggling to achieve the latest Quadruple Aim – reducing costs, and improving population health, patient experience, and team well-being – and productivity for the better performance and enhancement of the healthcare system (Arnetz et al., 2020). Thus, the incorporation of intelligent machines and improved computing power has become necessary for the healthcare sector to achieve its goals extensively.

The application of artificial intelligence, specifically machine learning (Solanki and Pandey, 2019; Ahuja et al., 2019; Tayal et al., 2020; Singh et al., 2020; Priyadarshini et al., 2019; Kaur et al., 2021) and its recently evolved subsets natural language processing (Singh and Solanki, 2016) and deep learning, has experienced a tremendous acceleration in almost all verticals. We define artificial intelligence as a smart science that empowers machines with human comprehension, intelligence, and logical thinking abilities to accomplish a task. The revolutionary AI algorithms and software combined with modern hardware technologies have recently expanded their medical sciences and healthcare sector applications. Figure 12.1 shows a graphical representation of the rise in the publications in PubMed from the year 2000 to 2020 with the search query: artificial intelligence and genomics or health.

12.2 THE DATA OVERLOAD

Data generation has drastically increased in recent years. In this era of data, every human generates a massive amount of data every day directly or indirectly. A digital

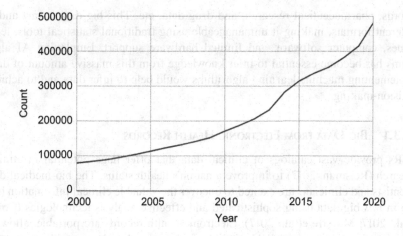

FIGURE 12.1 Graphical representation of rise in AI application research concerning publications in PubMed from the year 2000 to 2020 with the search query: artificial intelligence and genomics or health.

universe is a collection of all data created, cloned, and consumed. A recent report published by IDC and EMC Digital Universe ("The Digital Universe in 2020") projects that our digital universe will double its data every 2 years and reach 40 trillion GBs of data by 2020. That means every human would contribute approximately 5200 GB of data ("The Digital Universe in 2020: Big Data, Bigger Digital Shadows, and Biggest Growth in the Far East" by John Gantz and David Reinsel sponsored by EMC, n.d.). Though the amount of useful data is expanding and is being stored on the internet giants, like Google and other cloud service providers, only a tiny fraction of the world's Big digital Data potential is being realized.

The term "big data" refers to the massive data collection from the same or different origins. According to Douglas Laney, big data is governed by three Vs, namely volume, velocity, and variety, all growing in different dimensions. Laney's first V – volume – indicates the massive data generated. This big digital data has been generated rapidly over the past few years and is available for analysis which signifies the second V – velocity. The third V – variety – represents the diversity in big data such as video, audio, text or log files, images, etc.

12.3 BIG HEALTHCARE DATA

Over the past decade, a significant portion of the human population has realized the importance of data and is actively involved in generating it. The advent of smart technologies at lower prices has allowed us to generate and record healthcare data every day. These sharp sensors are highly sensitive and detect and record even minute alterations. These advancements serve as the basis of the global healthcare transformation by providing big multi-dimensional and heterogeneous data.

Our healthcare sector is data-rich. The major dimensions of healthcare data include EHRs, medical imaging, high-throughput genome sequencing, insurance records, pharmaceutical research, and drug data, etc. This big data is raw and in different formats, making it unmanageable using traditional statistical tools, techniques, computer software, and limited hardware support. Employing AI algorithms has become essential to infer knowledge from this massive amount of data. Implementing machine learning algorithms would help to infer data and to achieve decision-making.

12.3.1 BIG DATA FROM ELECTRONIC HEALTH RECORDS

EHRs provide vast catalogs of clinical data that offer huge potential in clinical research (Reisman, 2017) to improve a nation's health status. The bio-medical data scientists and clinicians are excited to uncover the valuable clinical information hidden in *this* big data using sophisticated and effective analysis technologies (Cowie et al., 2017; Meystre et al., 2017). Electronic health records are portable, allowing clinicians to reuse these available datasets for a longer duration, providing a recording of the patient's historical events of various healthcare parameters. EHR data is scattered and according to experts from the International Data Corporation, the unstructured data contributes approximately 80% of currently available healthcare data (Martin-Sanchez et al., 2014). Thus, it is challenging for EHR-based informatics researchers to make this data usable and interpretable.

12.3.2 BIG DATA FROM OMICS

The development of new techniques has enriched our healthcare sector with a large amount of data in a plethora of sub-branches like genomics, proteomics, metabolomics, etc., opening new avenues for personalized diagnostics and medicines. The last few decades have experienced a paradigm shift in high-throughput sequencing technologies. The advanced revolutionary techniques have accelerated the process of sequencing and have led to a drastic decline in the cost of sequencing. It is evident from the latest data provided by the National Institutes of Health that the cost per raw megabase of DNA sequence has declined from $5,292 in the year 2001 to $0.008 in the year 2020 (*DNA Sequencing Costs: Data*, n.d.) (Figure 12.2). Genomics data in the upcoming decade is expected to overcome other data-intensive disciplines including social media and YouTube. It is estimated that by the year 2025, the number of human genomes sequenced will probably reach 100 million to 2 billion (Stephens et al., 2015). This era of big biological data is creating new opportunities and challenges for researchers as well as organizations that work to store this data.

The complete characterization of different elements of a genome of an organism is called a genome project. One such example is the Human Genome Project (HGP), a collaborative effort of more than 2,800 researchers across the globe to generate complete, mapped, and annotated information of all human beings' genes. The first draft of the human genome with 3 billion base pairs (90% complete) was published in *Nature* in February 2001 by the International Human Genome Sequencing Consortium. Finally, the human genome project was accomplished covering 99% of the genome with 99.99% accuracy in April 2003 (Abdellah et al., 2004). The Human

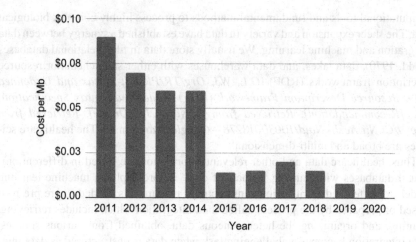

FIGURE 12.2   The decline in cost of sequencing from the year 2011 to 2020 as reported by the National Institutes of Health.

Genome Project's massive success motivated other initiatives like the International HapMap Project, 1000 Genomes, the Cancer Genome Atlas, Human Brain Project, and the emerging Human Proteome Project.

IndiGen is an excellent example of India's initiative to understand the Indian genome by conducting a pilot project study of more than 10,000 people of major ethnicities across India (Jain et al., 2021). The whole-genome sequence data so produced is highly recommended to understand better the genetic diseases, risk of cancer, and other clinical issues. An omics profile of an individual can be built by integrating all the omics data, namely, transcriptomics, metabolomics, proteomics, genomics, epigenomics, metabolomics, etc. Systematic and consolidative investigation of omics data strongly integrated with the healthcare records would help physicians to shape treatment strategies.

12.3.3   BIG DATA FROM MEDICAL IMAGES

Medical images are crucial for many disease diagnostics due to certain limitations of human vision and computing power. Diagnostic medical imaging is experiencing a transition from manual image analysis to highly sensitive and accurate AI-based techniques of image analysis. Before the AI-based analysis of the medical image data, approval from a local ethical committee is required. The approved data can then be stored, shared, accessed, and queried.

12.4   DATA PREPROCESS AND DATA INTEGRATION

The data we usually obtain from different sources is raw. Thus, it becomes necessary to preprocess the data and prepare it to generate knowledge from it. Data scientists usually prefer artificial intelligence and machine learning to develop novel computational tools and fundamental methods to process highly complex biological data. The sheer expansion and variety in data have established a synergy between data integration and machine learning. We usually store data in files, relational databases (Codd, 1970), data lakes, and data warehouses with either star schema or resource description frameworks (RDF) (*O.L. W3. Org/TR/PR-Rdf-Syntax and Undefined 1999. Resource Description Framework (RDF) Model and Syntax Specification, W3c [Recommendation]. Retrieved from Ci.Nii.Ac.Jp [Online]. Retrieved from Https://Ci.Nii.Ac.Jp/Naid/10030018278 – Google Search*, n.d.). The healthcare sciences are broad and multi-dimensional.

Thus, healthcare data and other relevant information are stored in different biological databases with varying formats of data. Before applying machine learning models, data fetched from various unstructured resources is needed to pre pre-processed or prepared by the process called data integration which includes retrieving, cleaning, and organizing the heterogeneous data, obtained from various sources. Data integration becomes a challenging task when data is unstructured as data integration is completely dependent on the data reading algorithms like downloading, organizing, and storing data. "Lazy" and "eager" frameworks for data integration are two different approaches commonly used for data integration (Lapatas et al., 2015). The lazy approach is when data is not accumulated at hub storage and stays at its original sources to be only called on-demand, interacting with their parent databases. On the other hand, in an eager approach, data is captured at a central storage hub like a warehouse structured according to its local schema. The non-uniformity in the data formats, their entities, and relationships make it inconvenient for the data scientist community to integrate such heterogeneous data and hinders the data integration process. Therefore, to overcome this problem, certain terms, scientific notations, and formats are defined and universally accepted to represent healthcare data and related terms. The following standards are the key to data integration and knowledge sharing.

12.5   DATA EXPLORATION

Data exploration is a critical step in data analysis. It is a technique of extracting information from data or getting insights into data. Performing data exploration adequately by using appropriate algorithms helps to identify novel critical information from big data. Earlier manual data analysis and human intuitions were generally used to analyze data but, with a big bang in the available data, it is not feasible to curate data without using computers. Researchers are now employing statistical tools and machine-based learning techniques to explore data. The two most popular and independent techniques of machine learning are supervised learning and unsupervised learning.

The supervised machine learning technology is governed by the training data to infer a general set of rules. The training data set for any supervised machine learning model is the table of instances, labeled with the class to which they belong. The quality of a supervised machine learning model completely depends on the quality and quantity of the training data set associated with the model.

The big health-related data provides researchers significant opportunities to employ artificial intelligence to decode an individual's genome for precision medicine or to predict the impact of a gene mutation. AI can accurately analyze a patient's medical reports and enable interventions specific to a patient by reading and analyzing massive clinical datasets much faster than a human. A recently released software platform based on ML to explore pharmaceutical partnerships is Dx: Revenue of Amplion (a leading precision medicine intelligence company). Fabric Genomics and Verge Genomics are AI-driven platforms enabling clinical labs to uncover critical and timely genomic insights that reduce costs and save lives. According to MarketsandMarkets' forecast, this industry, also called intelligent health or the smart hospital market, is expected to be valued at USD 63.49 billion by 2023, growing at a CAGR of 24.00% between 2017 and 2023 (*Smart Hospital Market | Industry Analysis and Market Size Forecast to 2023 | MarketsandMarkets™*, n.d.).

In order to extract hidden critical information, a range of software tools were developed to perform medical image analysis. For example, five different types of brain images (i.e. MRI, fMRI, PET, CT-scan, and EEG) can be easily analyzed using the SPM (Friston, 2003). ChesXpert is an amazing device developed by Stanford Group of Machine Learning which uses machine learning to detect pneumonia in less than ten seconds by analyzing the X-ray scans of patients (Rajpurkar et al., 2017). It is much faster, more accurate, and more efficient than conventional methods in detecting pneumonia. Big data analysis focused on bioinformatics can derive greater insights and value from imaging data to improve and sustain precision medicine programs, support resources for clinical decision-making, and other modes of healthcare.

12.5.1 ARTIFICIAL INTELLIGENCE IN CLINICAL DIAGNOSTICS

Artificial intelligence is computer programs derived from humans that are designed to analyze big data by performing human-like intelligent tasks. The ultimate objective of such tools is to produce models with high precision and speed. Major advances in AI algorithms have extended their use in the field of medical sciences, complemented by highly sophisticated processing units such as surgical robots (Dias & Torkamani, 2019), chatbots (Lu et al., 2016), smart devices such as smartwatches (Palanica et al., 2019), clinical diagnostics (Lee et al., 2018), and the new clinical diagnostics using genomics data. Artificial intelligence has proven its image-based diagnostics ability, but it is in its early stages when it comes to genomic-based clinical diagnostics. Different AI algorithms monitor various health problems. Computer vision techniques (Sundaram et al., 2018), for instance, address radiological X-ray image data in terms of numbers. The automated detection of COVID-19 by scanning X-ray images via neural networks (Rajkomar et al., 2018) is the most recent example. The use of speech recognition techniques (Esteva et al., 2019) detects neurological conditions. Time-series data analysis techniques are preferred for tracking variable data such as heartbeat and blood pressure (Fraser et al., 2015). The electronic health records-based medical industry relies on natural language processing techniques to capture useful keywords from either text or sound information and speech accordingly (Ozturk et al., 2020). In short, we can assume that diagnostic artificial intelligence is more specifically governed by deep learning or deep neural networks (Torkamani et al., 2017). Neural networks in AI are similar to our body's neural networks. Neurons (nodes in this case) exist in layers and transmit signals across all layers through interconnected nodes. All other layers are concealed, except the input layer and the output layer. The input data differs in correspondence with the data used to train the model for any application of a neural network. For example, pixel data must be fed to the neural network for picture analysis. Similarly, a neural network would prefer the high-throughput nucleotide sequences on which the model is trained (Zou et al., 2019) for diagnostics based on genomics evidence.

12.5.2 Artificial Intelligence in EHR-Based Diagnostics

The adoption of longitudinal EHRs has improved the availability of timely and accurate clinical data that may boost scores' predictive capacity if analyzed properly. Predictive decisions like clinical decision-making (Bell et al., 2014), diagnosis (Jolly et al., 2014), risk assessment, and treatment are extensively dependent on the accuracy and precision in analyzing these EHR records. Thus, it is challenging and critical to unlock the hidden information of these complex and heterogeneous data (Wright et al., 2015).

The application of artificial intelligence to EHR has been extensively used in various clinical studies like cardiology including the early detection of heart failure (Lopez-Jimenez et al., 2020), ophthalmology (Kern et al., 2020), predictive analysis of risk associated with cataract surgery, and risk assessment of diabetes (Hecht et al., 2020). The most commonly used AI technique to extract knowledge from unstructured language data (text/speech) is natural language processing (NLP). Figure 12.3 shows a typical NLP algorithm for text processing and patient-level disease classification.

12.5.3 Artificial Intelligence in Image-Based Clinical Diagnostics

Artificial intelligence has brought revolutions when it comes to image-based diagnostics. It has enhanced the accuracy and precision of diagnostics techniques and takes less time to make potential decisions, at a lesser cost. AI streamlines the process of analysis of any kind of medical imaging data like X-rays, CT-scans, magnetic resonance imaging (MRI), molecular imaging, ultrasound, photoacoustic imaging, molecular imaging technique, positron emission tomography (PET), electroencephalography (EEG, an electrophysiological monitoring method), and low X-ray-based mammograms for breast cancer detection. These imaging techniques generate high-definition medical images of patients. Though medical experts like radiologists and doctors efficiently analyze these image data to identify abnormalities, there are possibilities of missing out on critical information especially when the disease is emerging or the data is unstructured. Image processing affects healthcare by accurately identifying disease biomarkers from bio-medical images in order to aid in certain circumstances. To transform patients' diagnosis, treatment, and monitoring,

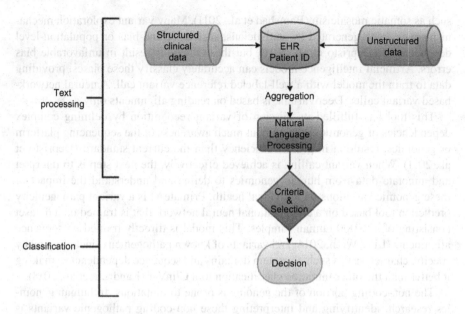

FIGURE 12.3 This shows a typical NLP algorithm for text processing and patient-level disease classification.

this method uses ML and pattern recognition techniques to analyze vast amounts of clinical image data. It focuses on improving the medical imaging diagnostic potential for clinical decision-making (Willemink et al., 2020).

Another computer vision-based application of AI (deep learning) is DeepGestalt (Gurovich et al., 2017). It is an image analysis and processing computer program that detects an individual's genetic defect syndrome through their facial features. Its well-trained model on large datasets governs the high accuracy of this tool.

12.5.4 ARTIFICIAL INTELLIGENCE IN GENOMICS-BASED CLINICAL DIAGNOSTICS

In the diagnosis of a disease such as cancer (Eraslan et al., 2019) mutation detection (Levy et al., 2007; Veltman & Brunner, 2012; Watson et al., 2013), and understanding personal genome sequencing (Campbell & Eichler, 2013) are popular these days. The use of standard statistical instruments and techniques in concatenation with human intelligence to evaluate high-throughput genomics sequence datasets has many drawbacks and is vulnerable to human errors. Applying AI makes the process of data interpretation quick and precise and decreases the risk of missing any minute observation.

The most sensitive and critical aspect of genome data analysis is detecting individual genomic variants (INDELS). In clinical diagnostics, this method is generally known as variant calling. Variant calling stages are susceptible to errors due to unwanted influence from sample preparation on the sequencing technologies used, background and interpretation of sequence, and even unpredictable biological factors such as somatic mosaicism (Bamshad et al., 2011). Many variant exploration mechanisms have been generated by statisticians, such as strand-bias or population-level dependencies (Depristo et al., 2011), but these methods result in unfavorable bias errors. Artificial intelligence models can accurately classify these biases, providing data to train the model with a well-labeled reference variant call. A neural network-based variant caller, DeepVariant, is based on reading alignments only.

This tool has fulfilled its mission of variant recognition by defining complex dependencies of genomics data without much awareness of the sequencing platform or genomics, resulting in greater efficiency than the current standard (Depristo et al., 2011). When variant calling is achieved effectively, the next step is to interpret and annotate data from human genomics to define and understand the impact of these genomic variations on individual health. PrimateAI is a variant pathogenicity prediction tool based on a computational neural network that is trained on a dataset consisting of 120,000 human samples. This model is directly trained on sequence alignments (Li & Wren, 2014) and variants of known pathogenicity, including many specific characteristics such as protein domains and sequence dependencies, making it better than the other mutation classification tool ClinVar (Landrum et al., 2018).

The non-coding portion of the genome is prone to mutations. In human genomics research, identifying and interpreting these non-coding pathogenic variants is a challenge (Sundaram et al., 2018). AI helps us to understand these non-coding variants. Gene splicing accounts for around 10% of pathogenic genomic variations (Chatterjee & Ahituv, 2017), but the many factors influencing gene-splicing make it difficult to detect these non-coding genetic variations (Jaganathan et al., 2019; Soemedi et al., 2017).

SpliceAI is a neural network-based method that has improved the accuracy of detecting non-coding variants from 57% to 95% by using 10 kb window size long-range sequence information (Baeza-Centurion et al., 2019). The input in the form of sequence data includes all the intronic and exonic regions. SpliceAI can predict the splicing events (both general and rarer events) by its multi-layer neural network model.

For clinical genome analysis, machine learning models are developed to identify genetic variants like somatic and copy number variants, which are usually not easy to detect. AI-based tools are also being developed for uncovering the impact of these genetic variants on downstream pathways and critical proteins associated with them in terms of measurement of gene expression data. To facilitate the clinical practices and decision-making in the healthcare and diagnostics domain, the International Business Machines Corporation (IBM), a renowned American information technology-based company with its headquarters in New York, developed IBM Watson Health (*IBM Watson Health | AI Healthcare Solutions – India | IBM*, n.d.) which uses artificial intelligence and data processing to assist physicians with clinical decision support systems.

12.6 MACHINE LEARNING IN CANCER PROGNOSIS

The application of machine learning techniques in disease detection has accelerated in the last few decades including cancer. Most of these supervised learning algorithms are the intersection of artificial intelligence and statistics to predict: (i) risk or susceptibility assessment, (ii) probability/chances of recovery, (iii) estimation of the rate of survival (Kourou et al., 2015).

The basic principle behind these machine learning-based techniques is the analysis of the large datasets from the patients along with the histological and clinical status, to model the progression of cancer and identify its unique characteristic features to be used in the classifier later.

In cancer research, artificial neural networks and decision trees have been used for decades to detect cancer. As per the latest cancer prognostics and predictions, machine learning models are now developed by integrating two or more machine learning algorithms like integrating support vector machines and decision trees, etc. These models are trained on heterogeneous data for the detection of tumors. As per the latest PubMed records, more than 16,000 research articles on the application of artificial intelligence in cancer can be easily accessed.

Earlier, the physicians were dependent on the clinical and histopathological data to make conclusions about a patient's disease status, say cancer detection/prognosis. The prediction of cancer usually means life expectancy, progression, treatment, and survivability of the patient. Integrating the clinical data with the macro-scale data like age, sex, weight, family history, habits, exposure to carcinogens, etc., assists the physicians in making decisions. But this information is nearly not enough to make robust predictions based on statistical tools and techniques.

Cancer predictions can be more accurate and precise if integrated with the data like gene expression data, molecular biomarkers, cellular parameters, etc. The recent advancements in high-throughput instrument technology bring all this highly descriptive information together, much faster and at a lesser cost. Despite the easy availability of such valuable data, it is quite a challenging task for physicians to analyze this high-throughput data, and thus they are dependent on intelligent machines and software applications to identify patterns and relationships between different entities in the complex datasets and make efficient decisions about the disease class.

Google's DeepMind is artificial intelligence-based smart technology. According to findings published in *Nature*, the Imperial College of London and Google Health researchers worked in collaboration to train a machine learning model with a training dataset of approximately 29,000 women to identify abnormalities in X-ray images with greater accuracy. Based on such applications, it is evident that integrating the heterogeneous data with smart learning algorithms has great potential to positively influence cancer detection and treatment (McKinney et al., 2020).

12.7 LIMITATIONS OF ARTIFICIAL INTELLIGENCE IN HEALTHCARE

Artificial intelligence algorithms have immense potential to identify critical relationships among different entities. But, when applied in the healthcare domain, AI is always susceptible to some risks. Thus, here we summarize the potential risks and challenges associated with medical AI specifically focusing on its limited use in diagnostics.

12.7.1 DATA DEPENDENCY AND INCONSISTENT DATA

Most AI algorithms are dependent on available data to make predictions. This dependence of AI on data limits its application in medical sciences as the healthcare data is huge, inconsistent, heterogeneous, unstructured, less annotated, and noisy. Thus, medical AI demands extensive data preprocessing steps and standardizing.

12.7.2 INFRASTRUCTURE REQUIREMENTS

Robust, scalable, and safe infrastructure in terms of quality data, digital storage, GPU technology, and other hardware and technology are considerable requirements of AI for high-performance computing.

12.7.3 DATA PRIVACY AND SECURITY

The preservation of privacy, confidentiality, and the ethical use of data is always an important concern while sharing healthcare data with AI commercial experts. Machine learning models are also prone to cyber-attacks. And cases of deliberate hacking of medical data can cause severe damage in this industry.

12.8 DISCUSSION

The healthcare industry across the globe is significantly dependent on health practitioners. Many countries face a challenge to maintain their people's health due to the severe scarcity of trained medical professionals. In such a scenario, medical AI-assisted bioinformatics is a boon. From medical records to diagnostics or patient care, AI has explicitly shown its potential through its different algorithms and models. Despite its extraordinary features, AI-based tools are not completely accepted especially when it comes to high-risk diagnostics. Many factors like inconsistent data and machine bias limit the efficiency of an AI model. We need to build more sophisticated and reliable intelligent tools based on hybrid AI models to expand medical care acceptance.

BIBLIOGRAPHY

Abdellah, Z., Ahmadi, A., Ahmed, S., Aimable, M., Ainscough, R., Almeida, J., Almond, C., Ambler, A., Ambrose, K., Ambrose, K., Andrew, R., Andrews, D., Andrews, N., Andrews, D., Apweiler, E., Arbery, H., Archer, B., Ash, G., Ashcroft, K., ... Kamholz, S. (2004). Finishing the euchromatic sequence of the human genome. *Nature*, *431*(7011), 931–945. https://doi.org/10.1038/nature03001

Ahuja, R., A. Solanki, & A. Nayyar (2019). Movie Recommender System Using K-Means Clustering AND K-Nearest Neighbor. in *2019 9th International Conference on Cloud Computing, Data Science & Engineering (Confluence)*. IEEE.

Arnetz, B. B., Goetz, C. M., Arnetz, J. E., Sudan, S., Vanschagen, J., Piersma, K., & Reyelts, F. (2020). Enhancing healthcare efficiency to achieve the quadruple aim: an exploratory study. *BMC Research Notes*, *13*(1), 362. https://doi.org/10.1186/s13104-020-05199-8

Baeza-Centurion, P., Miñana, B., Schmiedel, J. M., Valcárcel, J., & Lehner, B. (2019). Combinatorial genetics reveals a scaling law for the effects of mutations on splicing. *Cell, 176*(3), 549–563.e23. https://doi.org/10.1016/j.cell.2018.12.010

Bamshad, M. J., Ng, S. B., Bigham, A. W., Tabor, H. K., Emond, M. J., Nickerson, D. A., & Shendure, J. (2011). Exome sequencing as a tool for Mendelian disease gene discovery. In *Nature Reviews Genetics* (Vol. 12, Issue 11, pp. 745–755). Nature Publishing Group. https://doi.org/10.1038/nrg3031

Bell, G. C., Crews, K. R., Wilkinson, M. R., Haidar, C. E., Hicks, J. K., Baker, D. K., Kornegay, N. M., Yang, W., Cross, S. J., Howard, S. C., Freimuth, R. R., Evans, W. E., Broeckel, U., Relling, M. V., & Hoffman, J. M. (2014). Development and use of active clinical decision support for preemptive pharmacogenomics. *Journal of the American Medical Informatics Association, 21*(E2). https://doi.org/10.1136/amiajnl-2013-001993

Campbell, C. D., & Eichler, E. E. (2013). Properties and rates of germline mutations in humans. In *Trends in Genetics* (Vol. 29, Issue 10, pp. 575–584). Trends Genet. https://doi.org/10.1016/j.tig.2013.04.005

Chatterjee, S., & Ahituv, N. (2017). Gene regulatory elements, major drivers of human disease. *Annual Review of Genomics and Human Genetics, 18*(1), 45–63. https://doi.org/10.1146/annurev-genom-091416-035537

Codd, E. F. (1970). A relational model of data for large shared data banks. *Communications of the ACM, 13*(6), 377–387. https://doi.org/10.1145/362384.362685

Cowie, M. R., Blomster, J. I., Curtis, L. H., Duclaux, S., Ford, I., Fritz, F., Goldman, S., Janmohamed, S., Kreuzer, J., Leenay, M., Michel, A., Ong, S., Pell, J. P., Southworth, M. R., Stough, W. G., Thoenes, M., Zannad, F., & Zalewski, A. (2017). Electronic health records to facilitate clinical research. In *Clinical Research in Cardiology* (Vol. 106, Issue 1). Dr. Dietrich Steinkopff Verlag GmbH and Co. KG. https://doi.org/10.1007/s00392-016-1025-6

Depristo, M. A., Banks, E., Poplin, R., Garimella, K. V., Maguire, J. R., Hartl, C., Philippakis, A. A., Del Angel, G., Rivas, M. A., Hanna, M., McKenna, A., Fennell, T. J., Kernytsky, A. M., Sivachenko, A. Y., Cibulskis, K., Gabriel, S. B., Altshuler, D., & Daly, M. J. (2011). A framework for variation discovery and genotyping using next-generation DNA sequencing data. *Nature Genetics, 43*(5), 491–501. https://doi.org/10.1038/ng.806

Dias, R., & Torkamani, A. (2019). Artificial intelligence in clinical and genomic diagnostics. In *Genome Medicine* (Vol. 11, Issue 1, p. 70). BioMed Central Ltd. https://doi.org/10.1186/s13073-019-0689-8

DNA Sequencing Costs: Data. (n.d.). Retrieved February 20, 2021, from https://www.genome.gov/about-genomics/fact-sheets/DNA-Sequencing-Costs-Data

Eraslan, G., Avsec, Ž., Gagneur, J., & Theis, F. J. (2019). Deep learning: new computational modelling techniques for genomics. In *Nature Reviews Genetics* (Vol. 20, Issue 7, pp. 389–403). Nature Publishing Group. https://doi.org/10.1038/s41576-019-0122-6

Esteva, A., Robicquet, A., Ramsundar, B., Kuleshov, V., DePristo, M., Chou, K., Cui, C., Corrado, G., Thrun, S., & Dean, J. (2019). A guide to deep learning in Healthcare. In *Nature Medicine* (Vol. 25, Issue 1, pp. 24–29). Nature Publishing Group. https://doi.org/10.1038/s41591-018-0316-z

Fraser, K. C., Meltzer, J. A., & Rudzicz, F. (2015). Linguistic features identify Alzheimer's disease in narrative speech. *Journal of Alzheimer's Disease, 49*(2), 407–422. https://doi.org/10.3233/JAD-150520

Friston, K. J. (2003). Statistical parametric mapping. In *Neuroscience Databases* (pp. 237–250). Springer US. https://doi.org/10.1007/978-1-4615-1079-6_16

Gurovich, Y., Hanani, Y., Bar, O., Fleischer, N., Gelbman, D., Basel-Salmon, L., Krawitz, P., Kamphausen, S. B., Zenker, M., Bird, L. M., & Gripp, K. W. (2017). *DeepGestalt-Identifying Rare Genetic Syndromes Using Deep Learning.*

Hecht, I., Achiron, R., Bar, A., Munk, M. R., Huf, W., Burgansky-Eliash, Z., & Achiron, A. (2020). Development of "Predict ME," an online classifier to aid in differentiating diabetic macular edema from pseudophakic macular edema. *European Journal of Ophthalmology*, *30*(6), 1495–1498. https://doi.org/10.1177/1120672119865355

IBM Watson Health I AI Healthcare Solutions - India I IBM. (n.d.). Retrieved February 20, 2021, from https://www.ibm.com/in-en/watson-health

Jaganathan, K., Kyriazopoulou Panagiotopoulou, S., McRae, J. F., Darbandi, S. F., Knowles, D., Li, Y. I., Kosmicki, J. A., Arbelaez, J., Cui, W., Schwartz, G. B., Chow, E. D., Kanterakis, E., Gao, H., Kia, A., Batzoglou, S., Sanders, S. J., & Farh, K. K. H. (2019). Predicting splicing from primary sequence with deep learning. *Cell*, *176*(3), 535-548. e24. https://doi.org/10.1016/j.cell.2018.12.015

Jain, A., Bhoyar, R. C., Pandhare, K., Mishra, A., Sharma, D., Imran, M., Senthivel, V., Divakar, M. K., Rophina, M., Jolly, B., Batra, A., Sharma, S., Siwach, S., Jadhao, A. G., Palande, N. V., Jha, G. N., Ashrafi, N., Mishra, P. K., A K, V., ... Sivasubbu, S. (2021). IndiGenomes: a comprehensive resource of genetic variants from over 1000 Indian genomes. *Nucleic Acids Research*, *49*(D1), D1225–D1232. https://doi.org/10.1093/nar/gkaa923

Jolly, S. E., Navaneethan, S. D., Schold, J. D., Arrigain, S., Sharp, J. W., Jain, A. K., Schreiber, M. J., Simon, J. F., & Nally, J. V. (2014). Chronic kidney disease in an electronic health record problem list: Quality of care, ESRD, and mortality. *American Journal of Nephrology*, *39*(4), 288–296. https://doi.org/10.1159/000360306

Kaur, H., Singh, S. P., Bhatnagar, S. & Solanki, A. (2021). Intelligent smart home energy efficiency model using artificial intelligence and internet of things. *Artificial Intelligence to Solve Pervasive Internet of Things Issues*, 183–210.

Kern, C., Fu, D. J., Kortuem, K., Huemer, J., Barker, D., Davis, A., Balaskas, K., Keane, P. A., McKinnon, T., & Sim, D. A. (2020). Implementation of a cloud-based referral platform in ophthalmology: making telemedicine services a reality in eye care. *British Journal of Ophthalmology*, *104*(3), 312–317. https://doi.org/10.1136/bjophthalmol-2019-314161

Kourou, K., Exarchos, T. P., Exarchos, K. P., Karamouzis, M. V., & Fotiadis, D. I. (2015). Machine learning applications in cancer prognosis and prediction. In *Computational and Structural Biotechnology Journal* (Vol. 13, pp. 8–17). Elsevier. https://doi.org/10.1016/j.csbj.2014.11.005

Landrum, M. J., Lee, J. M., Benson, M., Brown, G. R., Chao, C., Chitipiralla, S., Gu, B., Hart, J., Hoffman, D., Jang, W., Karapetyan, K., Katz, K., Liu, C., Maddipatla, Z., Malheiro, A., McDaniel, K., Ovetsky, M., Riley, G., Zhou, G., ... Maglott, D. R. (2018). ClinVar: Improving access to variant interpretations and supporting evidence. *Nucleic Acids Research*, *46*(D1), D1062–D1067. https://doi.org/10.1093/nar/gkx1153

Lapatas, V., Stefanidakis, M., Jimenez, R. C., Via, A., & Schneider, M. V. (2015). Data integration in biological research: an overview. *Journal of Biological Research-Thessaloniki*, *22*(1), 9. https://doi.org/10.1186/s40709-015-0032-5

Lee, S. I., Celik, S., Logsdon, B. A., Lundberg, S. M., Martins, T. J., Oehler, V. G., Estey, E. H., Miller, C. P., Chien, S., Dai, J., Saxena, A., Blau, C. A., & Becker, P. S. (2018). A machine learning approach to integrate big data for precision medicine in acute myeloid leukemia. *Nature Communications*, *9*(1). https://doi.org/10.1038/s41467-017-02465-5

Levy, S., Sutton, G., Ng, P. C., Feuk, L., Halpern, A. L., Walenz, B. P., Axelrod, N., Huang, J., Kirkness, E. F., Denisov, G., Lin, Y., MacDonald, J. R., Pang, A. W. C., Shago, M., Stockwell, T. B., Tsiamouri, A., Bafna, V., Bansal, V., Kravitz, S. A., ... Venter, J. C. (2007). The diploid genome sequence of an individual human. *PLoS Biology*, *5*(10), 2113–2144. https://doi.org/10.1371/journal.pbio.0050254

Li, H., & Wren, J. (2014). Toward better understanding of artifacts in variant calling from high-coverage samples. In *Bioinformatics* (Vol. 30, Issue 20, pp. 2843–2851). Oxford University Press. https://doi.org/10.1093/bioinformatics/btu356

Lopez-Jimenez, F., Attia, Z., Arruda-Olson, A. M., Carter, R., Chareonthaitawee, P., Jouni, H., Kapa, S., Lerman, A., Luong, C., Medina-Inojosa, J. R., Noseworthy, P. A., Pellikka, P. A., Redfield, M. M., Roger, V. L., Sandhu, G. S., Senecal, C., & Friedman, P. A. (2020). Artificial intelligence in cardiology: present and future. In *Mayo Clinic Proceedings* (Vol. 95, Issue 5, pp. 1015–1039). Elsevier Ltd. https://doi.org/10.1016/j .mayocp.2020.01.038

Lu, T. C., Fu, C. M., Ma, M. H. M., Fang, C. C., & Turner, A. M. (2016). Healthcare applications of smart watches: A systematic review. *Applied Clinical Informatics*, 7(3), 850–869. https://doi.org/10.4338/ACI-2016-03-R-0042

Martin-Sanchez, F., informatics, K. V.-Y. of medical, & 2014, undefined. (n.d.). Big data in medicine is driving big changes. *Ncbi.Nlm.Nih.Gov*. Retrieved February 20, 2021, from https://www.ncbi.nlm.nih.gov/pmc/articles/PMC4287083/

McKinney, S. M., Sieniek, M., Godbole, V., Godwin, J., Antropova, N., Ashrafian, H., Back, T., Chesus, M., Corrado, G. C., Darzi, A., Etemadi, M., Garcia-Vicente, F., Gilbert, F. J., Halling-Brown, M., Hassabis, D., Jansen, S., Karthikesalingam, A., Kelly, C. J., King, D., … Shetty, S. (2020). International evaluation of an AI system for breast cancer screening. *Nature*, 577(7788), 89–94. https://doi.org/10.1038/s41586-019-1799-6

Meystre, S. M., Lovis, C., Bürkle, T., Tognola, G., Budrionis, A., & Lehmann, C. U. (2017). Clinical data reuse or secondary use: current status and potential future progress. In *Yearbook of Medical Informatics* (Vol. 26, Issue 1, pp. 38–52). Yearb Med Inform. https://doi.org/10.15265/IY-2017-007

O.L. w3. org/TR/PR-rdf-syntax and undefined 1999. Resource description framework (RDF) model and syntax specification, W3c [Recommendation]. Retrieved from *ci.nii.ac.jp* [Online]. Retrieved from *https://ci.nii.ac.jp/naid/10030018278* - Google Search. (n.d.). Retrieved February 20, 2021, from https://www.google.com/search...

Ozturk, T., Talo, M., Yildirim, E. A., Baloglu, U. B., Yildirim, O., & Rajendra Acharya, U. (2020). Automated detection of COVID-19 cases using deep neural networks with X-ray images. *Computers in Biology and Medicine*, 121, 103792. https://doi.org/10.1016 /j.compbiomed.2020.103792

Palanica, A., Flaschner, P., Thommandram, A., Li, M., & Fossat, Y. (2019). Physicians' perceptions of chatbots in health care: Cross-sectional web-based survey. *Journal of Medical Internet Research*, 21(4), e12887. https://doi.org/10.2196/12887

Priyadarshni, V. , Nayyar, A. Solanki. A. & Anuragi, A.(2019) Human Age Classification System Using K-NN Classifier, vol. 1075.

Rajkomar, A., Oren, E., Chen, K., Dai, A. M., Hajaj, N., Hardt, M., Liu, P. J., Liu, X., Marcus, J., Sun, M., Sundberg, P., Yee, H., Zhang, K., Zhang, Y., Flores, G., Duggan, G. E., Irvine, J., Le, Q., Litsch, K., … Dean, J. (2018). Scalable and accurate deep learning with electronic health records. In *arXiv (Vol. 1)*. arXiv. https://doi.org/10.1038/s41746-018-0029-1

Rajpurkar, P., Irvin, J., Zhu, K., Yang, B., Mehta, H., Duan, T., Ding, D., Bagul, A., Langlotz, C., Shpanskaya, K., Lungren, M. P., & Ng, A. Y. (2017). CheXNet: radiologist-level pneumonia detection on chest x-rays with deep learning. *ArXiv*. http://arxiv.org/abs/1711.05225.

Reisman, M. (2017). EHRs: the challenge of making electronic data usable and interoperable. *P and T*, 42(9), 572–575. https://www.ncbi.nlm.nih.gov/pmc/articles/PMC5565131/

Smart Hospital Market | Industry Analysis and Market Size Forecast to 2023 | MarketsandMarkets™. (n.d.). Retrieved February 20, 2021, from https://www.marketsa ndmarkets.com/Market-Reports/smart-hospital-market-29319948.html

Singh, G., & Solanki, A. (2016). An algorithm to transform natural language into sql queries for relational databases. *Selforganizology*, 3(3), 100–116.

Singh, T., Nayyar, A., and Solanki, A. (2020) "Multilingual opinion mining movie recommendation system using RNN", *First International Conference on Computing, Communications, and Cyber-Security*, Springer, Singapore.

Soemedi, R., Cygan, K. J., Rhine, C. L., Wang, J., Bulacan, C., Yang, J., Bayrak-Toydemir, P., McDonald, J., & Fairbrother, W. G. (2017). Pathogenic variants that alter protein code often disrupt splicing. *Nature Genetics 49*(6), 848–855. https://doi.org/10.1038/ng.3837

Solanki, A.; Pandey, S. (2019). Music instrument recognition using deep convolutional neural networks. *Int. J. Inf. Technol. (IJITEE)* vol. 8, 1076–1079.

Stephens, Z. D., Lee, S. Y., Faghri, F., Campbell, R. H., Zhai, C., Efron, M. J., Iyer, R., Schatz, M. C., Sinha, S., & Robinson, G. E. (2015). Big data: astronomical or genomical? *PLOS Biology 13*(7), e1002195. https://doi.org/10.1371/journal.pbio.1002195

Sundaram, L., Gao, H., Padigepati, S. R., McRae, J. F., Li, Y., Kosmicki, J. A., Fritzilas, N., Hakenberg, J., Dutta, A., Shon, J., Xu, J., Batzloglou, S., Li, X., & Farh, K. K. H. (2018). Predicting the clinical impact of human mutation with deep neural networks. *Nature Genetics, 50*(8), 1161–1170. https://doi.org/10.1038/s41588-018-0167-z

The Digital Universe in 2020: Big Data, Bigger Digital Shadows, and Biggest Growth in the Far East by John Gantz and David Reinsel sponsored by EMC. (n.d.). Retrieved February 20, 2021, from https://www.emc.com/leadership/digital-universe/2012iview/index.htm

Tayal A., Kose U., Solanki A., Nayyar A., and Saucedo J. A. M., (2020). Efficiency analysis for stochastic dynamic facility layout problem using metaheuristic, data envelopment analysis and machine learning. *Comput. Intell.*, vol. 36, no. 1, pp. 172–202.

Torkamani, A., Andersen, K. G., Steinhubl, S. R., & Topol, E. J. (2017). High-definition medicine. In *Cell* (Vol. 170, Issue 5, pp. 828–843). Cell Press. https://doi.org/10.1016/j.cell.2017.08.007

Veltman, J. A., & Brunner, H. G. (2012). De novo mutations in human genetic disease. *Nature Reviews Genetics, 13*(8), 565–575.

Watson, I. R., Takahashi, K., Futreal, P. A., & Chin, L. (2013). Emerging patterns of somatic mutations in cancer. In *Nature Reviews Genetics* (Vol. 14, Issue 10, pp. 703–718). Nat Rev Genet. https://doi.org/10.1038/nrg3539

Willemink, M. J., Koszek, W. A., Hardell, C., Wu, J., Fleischmann, D., Harvey, H., Folio, L. R., Summers, R. M., Rubin, D. L., & Lungren, M. P. (2020). Preparing medical imaging data for machine learning. In *Radiology* (Vol. 295, Issue 1, pp. 4–15). Radiological Society of North America Inc. https://doi.org/10.1148/radiol.2020192224

Wright, A., Mccoy, A. B., Hickman, T.-T. T., St Hilaire, D., Borbolla, D., Bowes, L., Dixon, W. G., Dorr, D. A., Krall, M., Malholtra, S., Bates, D. W., & Sittig, D. F. (2015). Problem list completeness in electronic health records: a multi-site study and assessment of success factors HHS Public Access. *Int J Med Inform 84*(10), 784–790. https://doi.org/10.1016/j.ijmedinf.2015.06.011

Zou, J., Huss, M., Abid, A., Mohammadi, P., Torkamani, A., & Telenti, A. (2019). A primer on deep learning in genomics. *Nature Genetics 51*(1), 12–18. https://doi.org/10.1038/s41588-018-0295-5

13 Accelerating Translational Medical Research by Leveraging Artificial Intelligence
Digital Healthcare

*G. M. Roopa, K. Shryavani, N. Pradeep,
and Vicente Garcia Diaz*

CONTENTS

13.1 INTRODUCTION

From the recent past, it is observed that healthcare systems aim to cure diseases, elongate our lives, and enhance the patient community's comfort. Existing healthcare services would be benefited from the transformation in the usage of secure/

DOI: 10.1201/9781003126164-13

safe information technology support. This certifies the protection of health records, available when and where it is required, contributing to safer, better quality, more organized, and more productive and less expensive treatment for everyone [2].

But over the past three decades, the traditional clinical trial has remained mostly unchanged, which is partly due to uncertainty in regulatory needs, possibility of catastrophic failure. There is skepticism about quickly emerging technology and still untested technologies such as machine learning, wireless health tracking devices and sensors. There is a lack of strong actionable biomedical data sources and sophisticated analytics to create hypotheses that could inspire innovative diagnostics growth [6]. New techniques will also be needed to evaluate new biomedical approaches for safety and effectiveness, as it has been shown that current therapies sometimes only function for concise healthcare data. However, the use of new digital technology, such as next-generation sequencing, has improved both our knowledge of the disease mechanisms within broader patient populations and the potential for personalized therapies to be developed.

The consequences in most traditional clinical trials of ordinary treatment effects that do not readily help to make individualized, efficient clinical choices at the standard point of service are a crucial, daunting challenge in the clinical development process [8]. More streamlined procedures are promising approaches to overcoming this obstacle, leveraging new automated clinical endpoints and therapeutic reaction biomarkers capable of close and precise monitoring, improving efficacy and safety while minimizing toxicity and adverse reactions, and providing better insight on patient journeys by sensors and low-cost care. Protecting, standardizing, and improving regularly gathered healthcare data as a source of reliable clinical evidence promotes the organization of point-of-care clinical trials and can help to enhance the process of clinical growth [11].

The concept of A.I./M.L. is proposed to improve health services and to address the drawbacks in the traditional clinical trials. But from the recent past, A.I. has so far been inter-laced into everyday life with its use for standard search engines, spam email, and malware filters, and to uncover fake credit cards; A.I. directs individuals' demands in business realms, technology, and entertainment. Unfortunately, medical research has not fully adaptable to this revolution, with a limited number of A.I.-related medical applications available.

Artificial intelligence: Computing algorithms and frameworks that execute various tasks related to human intelligence that include decision systems, speech recognition, visual perception, and reasoning. A.I. involves diverse methods like machine learning, natural language processing, and computer vision.

Biomedical engineering: Presents a set of applications with analytical tools and engineering skills for bringing innovations in healthcare services and biology by promoting tools and devices adopted in clinical practices, therapeutics, and the physiological modeling of physical benefits.

Clinic: Any medical setup where patients are diagnosed and in particular counseled on health-relevant situations and is synonymously used for hospitals and health centers.

Machine learning: Beyond explicit operations consideration, model dependability, and inferences, the research study of algorithms/statistical models used by computer systems efficiently accomplishes an acceptable task.

In this context today, the adoption of artificial intelligence/machine learning has revamped the delivery of healthcare services that impact millions of patients. These disruptive technologies have transformed the way in which clinicians aid patients, enhance the treatment procedures, and carry out low-cost laboratory diagnostics. With learning space from such enormous data and exploring knowledge from unique occurrences by examining the key information from the health data that allows A.I. to move toward precise and personalized treatment, A.I. has an impact on influencing and reinforcing the conventional practices.

Current disruptive computing technologies like artificial intelligence and machine learning have changed the delivery of healthcare services by transforming the way clinicians support the patients with the enhancement of scientists and laboratory diagnostics. Now, academic medical researchers and biomedical companies regularly adopt high-end technologies like parallel sequencing, microscopic imaging, and compound screening with the fast expansion in the quality/volume of generated data. Thus, with the advanced M.L. techniques, it is suggested that the increase in the biomedical insights of Big-Data adapting to the illness procedures for the discovery of innovative treatment schemes to improve diagnostic tools connected to medical applications. Figure 13.1 reveals the set of disruptive technologies supported for the "Digital Healthcare Ecosystem."

To summarize: The critical priorities explored and identified for translation research which are addressed in this chapter are:

- Defining and highlighting the clinical challenges and issues that can be solved by adopting A.I. concepts.

FIGURE 13.1   Disruptive technologies support for the digital healthcare ecosystem.

- An overview of the efficient A.I. algorithms that promote generalizability to broader medical practices and to overcome the problems related to healthcare.
- Supporting tools that aid in the validation and monitoring of performance by the adaptation of A.I. algorithms.
- Strategies for upgrading/enhancing the clinical development process by the combination of A.I.-M.L.-based digital methods and computing tools to ensure and improve patient treatment.

In this chapter, the authors have focused on translational research to cater to the multidirectional and multidisciplinary blending of basic patient-oriented research. Support for Industry 4.0 with the long-term goal of improving public health.

The chapter mainly aims to provide an overview of the advances in artificial intelligence, machine learning, and robotics/automation systems to address the inter-related challenges related to healthcare systems and the problems for laboratory scientists due to massive incoming data records. Next, the chapter highlights how the increased computing support in parallel with the application of current technologies can help to increase the success of translation studies for medical research. Further, the chapter emphasizes the major transformation in clinical development by converging massive digital data resources and parallel processing resources to recognize the actual data patterns by exploring and applying relevant artificial intelligence and machine learning algorithms accordingly.

The chapter includes an introduction, background theory exploring the drawbacks of the existing approaches, challenges and motivation, and secure computing technologies to improve the medical care of patients. There is an overview of the efficient A.I. medical-practice algorithms, techniques for modernizing the clinical development process by incorporating A.I.-M.L.-based automated approaches, some shortcomings, and references.

13.1.1 ORIGIN OF ARTIFICIAL INTELLIGENCE

Artificial intelligence (A.I.) is a young subject of 60 years that includes a set of sciences, hypotheses, and techniques (like statistics, mathematical logic, computational neurobiology, probabilities, and computer science) that aims to emulate the cognitive potential of humans. Initiated during the Second World War, its development is closely linked to the set of computing programs. It has led computers to perform increasingly complex tasks that could previously be delegated to humans only.

However, this automation differs from human intelligence in some restricted sense, which presents it to open criticism by a few experts. The ultimate aim of their research (a "strong A.I.," the potential to contextualize various specialized complex problems fully in an autonomous way) is entirely not comparable to existing achievements ("weak/moderate A.I.s," immensely efficient in the training model). The "strong-A.I.," which has been imagined in science fiction, would need advancements in primary research (not just performance improvements) but has the ability to model the whole world.

The association between the area of medical science and A.I. has noticed a rapid evolution of new algorithms and computational procedures based on novel awareness of translational medical research. It is studied that, the human brain is crucial for implementing A.I. algorithms. With the study of humans' cognitive potential and its neural execution, researchers can look into various facets of higher standards of contextual intelligence. One of the significant advantages in exploring the human brain for the implementation of A.I. is that the human mind acts as a primitive fundamental unit for the performance of novel methods that are different from the mathematical-related A.I. development.

By imitating aspects of the human brain, as they understand it, scientists have tried to design/implement algorithms and computational procedures to execute specific tasks. Besides, the human brain can also evaluate a few existing developments of A.I. techniques. Various algorithms and computational programs have been proposed which focus on human mind, which forms the core segment of the model. Thus, the study of the brain can link to A.I. research, also intimating the association of social capital and finances. Algorithms and applications related to brain can lead to more awareness of A.I. implementation.

1. Role of A.I./M.L. in Translation Medicine

The massive expansions of Internet services and recent developments in high-throughput technology have made it easy for the public, particularly for the scientific community, to access critical biological datasets. As a result, in the recent past methods adopted to process, interpret, and infer information, whether with clinical/sequencing data, electronic health reports, or medicine, in general, have changed dramatically. Owing to this, in one way or another today computer science terms such as machine learning and artificial intelligence are part of our everyday language usage. How translational research is modeled and executed has already been revolutionized, leading to the applications to improve human health around the world.

Artificial intelligence and machine learning have started to transform the way we deliver healthcare. These innovative computational methodologies will transform the way clinicians support patients and will also improve the way clinicians produce novel laboratory therapy and diagnostics.

2. A.I./M.L. and Translational Research

Machine learning implementations include, but are not limited to, administrative streaming tasks, maintaining patient records, clinical treatments, and other diseases, etc. Figure 13.2 shows a few examples of the activities of machine learning as a part of A.I. algorithms with its significant support for medical science and healthcare.

The usage of A.I. and M.L. has been explored widely by conceptual research laboratories, biotechnology firms, and companies in major areas:

   i. Machine-oriented learning for predicting the medicinal characteristics of
      molecular composites and goals for efficient translation research activities.

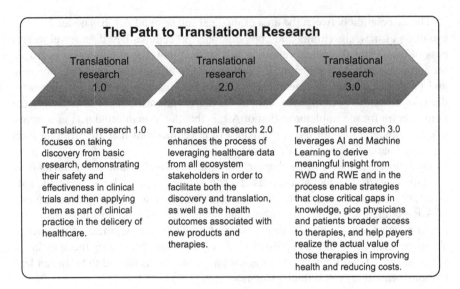

FIGURE 13.2   Translational research path (original).

ii. Using medical image pattern recognition and segmentation techniques to allow for quicker detection and monitoring of disease progression and generative A.I. algorithms to compute the growth of existing clinical and medical datasets.

iii. Developing deep learning techniques to detect new predictive models on inter-modal data sources, such as integrating genomic and medical data.

13.1.2   A.I. AND BIG DATA

Big data in medical care includes billions of data entries about treatments, patients, surgical procedures, research results, drugs, and much more. The use of such stored data was one of the key factors which led to the production and efficient validation of A.I. techniques. Observational experiments, in particular on anonymized medical records, indicate that gaining ethical consent may theoretically be more transparent than active research involving patients. This because accumulated data is usable and can be quickly used to verify an algorithm design while admitting new cases would prolong the algorithm's evaluation. Both physicians and researchers have recognized the utility of computing models in the processing and comprehension of preserved information.

There's an alternate association between big data and A.I., where the latter relies heavily on the former to gain success, while also helps organizations to unlock the potential of data stores in a manner that was previously complicated or impossible. However, big data in clinical care is rigidly linked to artificial intelligence in medical care where big data acts as an inception for artificial intelligence and powers up artificial intelligence to work efficiently. A.I. systems are furnished with crucial clinical information to aid in decision making that makes precise clinical predictions in no time and further minimizes the diagnostic and therapeutic errors. Future clinicians will rely, in their tasks and outcomes, almost absolutely on artificial intelligence. From the recent past, A.I. and big data operate together where all intelligent algorithms depend on a heap of data to draw precise decisions which gives rise to big data that provides the framework for developing and deploying more efficient and effective A.I.-based systems. In clinical terms, A.I. analyzes data from an extensive clinical study in a specific research context and predicts the effectiveness of particular treatments and procedures accurately. It also isolates data concerning patients' health status, ages, and other critical details. Today, it is seen that massive data drives better insights on clinical issues that need to be solved, but it is reviewed that the more data is applied to machine learning models, the better predictions are drawn.

A.I. and big data offer various efficient services for patients, clinicians, and researchers:

- Encouraging patient self-service with chatbots.
- Quicker computer-aided interface for diagnosis.
- Outlining image evidence to investigate the clinical data and to assess and monitor disease through automated service.
- More comprehensive healthcare data for personalized therapies.

13.1.3   OPTIMIZING THE MACHINE–HUMAN INTERFACE

A massive amount of theoretical knowledge derived from the digital electronic medical records during the observation is gradually adding more difficulty in understanding and interpreting such data. The blending of digital medical information from several sources, including medical imaging, laboratory, pathology, radionics, and genomics, is crucial for universal improved service tools with all aspects of medical care, which requires smashing down data silos so that a coherently combined resource bio-network is designed. And a work medium is developed that drives various physician experts together and integrates human intuition, and evaluation with A.I. results in boosting human skills to import personalized diagnosis into regular clinical routines. The availability of data silos could be achieved with the support of A.P.I. tools that provide total elegant with A.I. algorithms that efficiently mine the data from diverse resources for patient-related information. The medical academy has catalyzed this machine–human hybrid-system as the "Future Diagnostic Cockpit."

This complete integrated service will acquire standard digitized inputs from pathology, imaging, "omics," and other diverse clinical information retrieved from e-healthcare records and will store this in a data repository center available to clinicians throughout the patients' treatment. Clinical data from the repository is processed and rendered to the lab technician for the generation of the diagnostic report.

A.I. approaches are reliable enough to be efficient tools that render quantitative results to the pathologists, radiologists, and future medical experts with enhanced human–machine interfaces.

13.1.4 ROLE OF ARTIFICIAL INTELLIGENCE IN CLINICAL RESEARCH

In recent years, exposure to patient health records has dramatically altered many aspects of healthcare, from early discovery and testing to patient management, along with rapid advances in data collection techniques and technology. Advanced clinical practice is one of the most critical uses of modern technologies. For pharmaceutical firms, clinicians, consumers, policymakers, emerging technology, including artificial intelligence, offers the possibility of solving many of the most daunting facets of clinical challenges. The most promising future applications of artificial intelligence and machine learning in healthcare can be listed as:

1. Disease Identification and Diagnosis

    Recognizing and diagnosing diseases with machine learning present a framework for the computational exploration of outlines and the purpose of knowledge that allows health professionals to share adapted treatment known as personalized medicine.

2. Personalized Therapy/Behavioral Variations

    Personalized therapy is now a trendy research field and is closely connected to improved risk assessment, or more successful care based on human patient data combined with advanced predictive analytics. Supervised learning actually rules the domain, enabling doctors to pick from more restricted ranges of diagnoses, for example, or to predict patient risk depending on conditions and genetic details. Behavioral intervention is also an important factor of the prevention system.

3. Drug Discovery and Manufacturing

    From the initial test of new drugs to the expected success rate based on biological variables, the application of machine learning in initial drug development has the potential for different applications. This entails innovations for R&D discovery, such as next-generation screening. Personalized therapy appears to be the pioneer of this space, which requires the discovery of causes for "multi-factorial" disorders and, in turn, alternate routes for therapy. Most of this study requires unsupervised learning, which is also largely limited to the detection of data structures without forecasts.

4. Clinical Trial Research

    In focused clinical trial studies, machine learning has many valuable possible applications. Applying an advanced analytical framework to predict candidates for clinical trials could rely on a far broader variety of evidence, including social media and doctor visits, as well as genetic knowledge when targeting certain populations, resulting in fewer, easier, and less costly trials overall.

    For improved protection, M.L. may also be used for remote control and real-time data accessibility, such as tracking biological and other signs of any damage to or death of participants.

5. **Smart Health Records**

    In order to advance the processing and digitization of electronic medical records, report classification (for instance, sorting patient requests by email) using optical character recognition (transforming cursive or other sketched writings into computerized characters) and support vector machines are also important M.L.-based technologies.

6. **Influenza Pandemic Prediction**

   Based on information obtained from satellites, historical site information, real-time social networking site alerts, and other outlets, M.L. and A.I. technologies are also applied to tracking and forecasting disease outbreaks around the world. The drug crisis is a clear case of today's utilization of A.I. technology. For example, in order to forecast malaria outbreaks, taking into consideration information such as temperature, monthly average rainfall, the overall number of positive cases, and other data points, support vector machines and artificial neural networks have been used.

13.1.5   THE CORE ELEMENTS OF SMART HEALTHCARE COMMUNITIES (S.H.C.)

Today, groups of non-traditional participants, including government, non-profit, and corporate entities, have developed SHC-aggregated types of people focused on preventing disease and wellbeing, collaborating consistently, and functioning entirely beyond the conventional healthcare system. Communities can be regional or virtual, parts of the population (seniors, opioid users, mental health patients), self-trained or professional (weight management), or existing communities (schools, employers). Figure 13.3 shows the core components involved in the Smart Health Care communities.

The fundamentals of SHCs have been in use for many years, for example diabetes control, weight loss, and community-based addiction services that empower people to control their weight and exercise through a professional mentor proactively. Although successful, these initiatives have had limited scope because they are connected to building locations and require people to take part in person. Digital innovations can carry these services to a large degree, thereby growing their effects. SHC systems will use data-driven technology such as the Internet of Things (IoT) and augmented/virtual worlds to provide rapid and reliable decisions. The interaction model can be

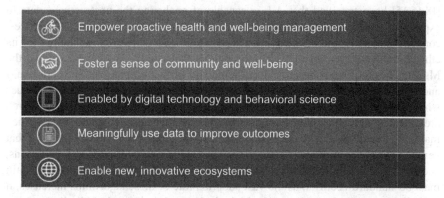

FIGURE 13.3   Core elements of S.H.C. (original).

adapted to each group for person-oriented, educational, and community-based support, and made possible by modular technologies. It may also incorporate an impact on health research, achieving evidence-based results, and drastically altering results.

The popularity of smartphones, which are widespread even in underprivileged communities, may improve the capability for virtual SHC services to be broadly scaled. Applications on these phones may integrate techniques from behavioral psychology, such as nudges and gamification, to help users keep track of their healthcare priorities.

There are difficulties in developing, maintaining, and sustaining successful SHCs which involve creating a business model structure for network partners and, at the same time, allowing knowledge exchange and patient privacy. It also tends to be much easier to recruit and maintain "wealthy, nervous, and happy" people than others who lack the incentive or ability to better their wellbeing. SHCs should implement methods that support the weak, the depressed, and the unhealthy to develop healthier habits, using behavioral modification research and programs that resolve mental health drivers. Despite these obstacles, SHC models can radically reshape how parts of the community work together to drive healthy societies.

13.2   RELATED STUDY

Authors in [1, 3] have discussed the existence of artificial intelligence in clinical practice as a significant unit in medical research that has quickly evolved with other modern fields like genomics, precision medicine, and tele consultation. The studies in [2, 6, 7] have demonstrated how the scientific progress should remain rigorous and opaque in providing novel solutions to enhance the current healthcare services, and clinical policies must focus on tackling the financial/ethical issues related to the corner-stone of medical evolution. The studies in [10, 9, 11] have presented through their review that A.I. has evolved as an efficient tool in medicine scenarios. With the familiarization of machine power and big data the basics of clinical practices and research will change. The studies in [12,13,14] have explored that conventional statistics are significantly high in a simple dataset and to evaluate that an area of clinical practices leads to robust predictions and to explore big data with A.I. The studies in [18,16,15] have specified that A.I. is capable of analyzing the unstructured data to expand medical research and besides, A.I. further supports for mobile health services, computing models, and the generation of synthetic data with novel regulations for its ethical and legal issues. In [19, 20], throughout their study, they have discussed the M.L. applications for translational medicine and further emphasized the core issues and limitations of these applications and the prerequisites to overcome these issues when adopted for clinical setups. Although M.L. applications to translational research are accepted with great interest, some measures should be used to reduce the risk they may pose to patients due to their adoption, and the authors in [4] have suggested that new lab practices in medicine use cases must adhere to peer-revised publications, and strict real-time validation and testing of A.I./M.L. algorithms used in the domain must be standardized. But, as the vast adoption rate of these algorithms increases, it is explicitly required to consider the biological questions instead of blindly applying these algorithms on the large volume of datasets. The authors in [9] have determined the critical clinical requirements that can be solved by A.I.-based M.L. algorithms for medical research. The priority for A.I. use cases is to determine if an algorithm can provide cost-effective analysis and make precise predictions.

13.2.1 PROBLEM STATEMENT

Currently, substantial e-healthcare records/reports are manually examined, and then the condition of the patients is decided which in turn becomes a time-consuming process. There is no consistency in human judgment and the diagnosis usually changes from human to human and even from day to day by the same individual. Today, clinical trials are growing increasingly complex, where the present research with the adoption of A.I. should enhance patient care. In order to achieve this, clinical trials must test results that represent real-world settings and concerns. These issues are pointed out by the "high rate of failure to reach primary endpoints due to bad or complicated architecture" [10].

This contradiction occurs because decision-makers usually lack understanding of the data records, which in turn leads to poor decision making. Moreover an insufficient period of follow-up and early-stop trials often lead to a shortage of credible data for decision-makers. Problems of missing data exist in nearly all studies which reduces analysis strength and can quickly lead to false conclusions. It is also important to specify the exact meanings for the outcomes of the trial since results that are not clearly defined can lead to uncertainty. When resolved, the questions of how the findings are selected, recorded, documented, and ultimately analyzed will continue to make a substantial contribution to the reasons why conventional clinical trial effects frequently fail to transfer into clinical benefit for patients.

13.2.2 TECHNOLOGY SUPPORT: PROPOSED SOLUTION

Today, technology is playing a significant role in enhancing various facets of clinical practices and is currently viewed as one of the big hopes in the coming future. However, in this regard, there is an entire heap of new challenges facing those participating in, and balancing clinical practices. From one of the witnesses: "with technology, massive changes are happening so rapidly, the challenge mainly deals with the selection and adoption of the appropriate technology, and for that to happen technology must gather the patient's acceptance, medical professionals, and regulators[5]." In this context, the adoption of artificial intelligence/machine learning for translation research nurtures the multidiscipline integration of medical research and patient-related research with the long-term goal of enhancing public health. It continues to promote the relationship between computing services, laboratory-focused research, and patient-related research in order to modularize a strong scientific understanding of patient care and illness.

Finally, to conclude on the technology support, interlinking with disciplines via multidisciplinary groups aids the advent of novel techniques and approaches to addressing the primary patient-related issues. The initiation and implementation of novel concepts are targets of scientific research and the adoption of artificial intelligence in clinics to transform existing modes of healthcare delivery. However, one should keep in mind the ethical and regulatory issues to avoid unnecessary risks and pitfalls that can hinder the "computers to clinics" flow. Figure 13.4 shows the working procedure adopted in traditional clinical research which initially formulates a hypothesis and then verifies with statistical analysis, whereas clinical research with A.I. is free from hypothesis/data-driven. The main scope of A.I. is to predict new instances and explore the hidden data patterns.

13.2.3 Research Challenges/Gaps and Infrastructure Requirements

As with other emerging innovations that have been translated from initial testing into widespread medical practice, it is important to recognize that there are significant barriers to the clinical application of A.I. tools (Figure 13.5). In order to recognize the essence of these emerging threats, effective mitigation methods and a well-designed study framework are needed to ensure success in A.I. The production of algorithms is ultimately translated to clinical research and is of the utmost priority.

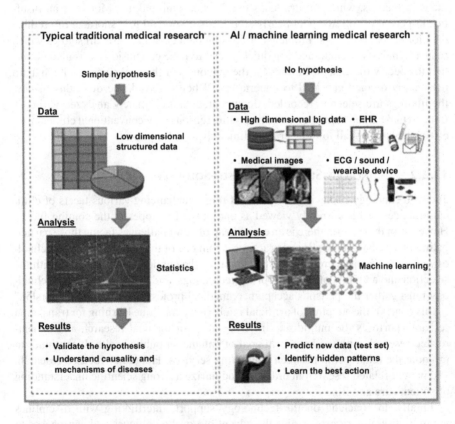

FIGURE 13.4 Clinical research pipeline with traditional practices and A.I.

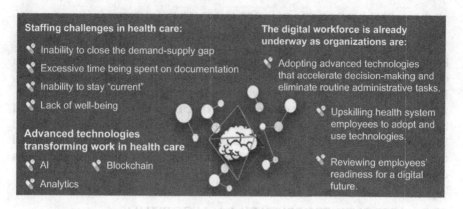

FIGURE 13.5 Challenges and research gaps with infrastructural needs.

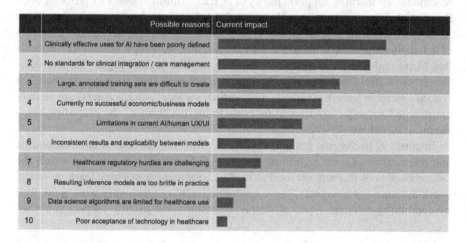

FIGURE 13.6 Lack of available data for training, testing, and validating A.I. algorithms.

More work in A.I. design is being carried out in single organizations with their data for the preparation, testing, and evaluation of A.I. algorithms. A recent report of the study investigating the efficiency of A.I. algorithms for therapeutic disease research showed that only 6 per cent of the 516 reviews analyzed had done external validation. So far, preliminary research has shown that these algorithms are generalizable to mainstream clinical practice.

From Figure 13.6, it is seen that while initial implementations of A.I. user interface in healthcare are interesting, there is limited use of A.I. in daily medical services. From the problems posed by researchers and engineers to the application of A.I. in clinical practice, there is lack of organized data for the planning, assessment, and examination of A.I algorithms and the implementation of A.I. procedures applied on patient treatment situations (UI = user interface and UX = user experience).

13.3 METHODOLOGY: PHASES INVOLVED IN THE ADOPTION OF A.I. FOR TRANSLATION RESEARCH

Figure 13.7 describes the standard methodology adopted for the implementation of A.I. in clinical trials with a registry report to document real-world outcomes. Information on algorithm complexity is obtained by a traditional reporting framework, and metadata on the test parameters is extracted from the context and retrieved by different registries, including A.I. monitoring of test registries. Reports should be given to all concerned, including the radiologist, the algorithm developer, and the regulatory authorities, and, finally, inferences should be taken from the results obtained.

13.4 APPROVED PROPRIETARY A.I. ALGORITHMS

Machine/deep learning supports plenty of A.I. algorithms. Table 13.1 provides brief descriptions of the most commonly adopted A.I.-based machine learning algorithms for several real-time tasks. Currently, ensemble/deep learning acts as the core part of A.I. algorithms. Ensemble techniques are machine learning conventional methods that blend multiple "weak-learners" (algorithms) like logistic regression and decision trees to attain better prediction. Bagging, boosting, and stacking are the three crucial ensemble learning methods. In boosting, various "weak learners" are serially merged and subsequently trained by considering the flaws of the preceding algorithms to minimize skewness.

In contrast, in bagging multiple "weak learners" are introduced simultaneously, and the outputs of each algorithm are stacked. Stacking is method by which the outcomes of poor learners are used as feedback for another machine learning model (meta-learning environment). These ensemble learning techniques work quite well by combining different types of simple algorithms and usually outperform when

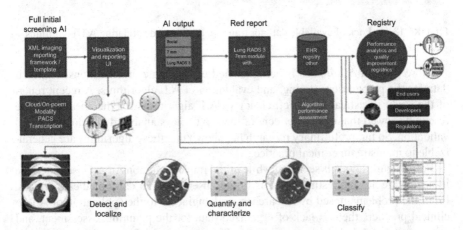

FIGURE 13.7   Standard methodology adopted for A.I.-based translation research.

TABLE 13.1

Commonly Adopted A.I.-Based Machine Learning Algorithms

Algorithm	Description	Use
"Logistic regression"	"Determines the likelihood of a contextualized endpoint with linear model"	"Classification"
"Decision tree"	"Separates data into branches via assessing knowledge gain and represents algorithm output (class/value)"	"Classification/ regression"
"Neural network"	"Relates to human brain design with layers made up of interconnected blocks with weighted edges according to the training outcomes"	"Classification/ regression"
"K-nearest neighbor"	"Classifies results by considering the k instances that occur in the method"	"Classification/ regression"
"Support vector machine"	"Produces a boundary that maximizes the distance of each class""	"Classification/ regression"
"K means"	"Allows k cluster nodes where each observation fits into the closest observation positions""	"Clustering"
"Hierarchical clustering"	"Builds a dendrogram with cluster hierarchy. Clusters pairs are combined to generate clusters""	"Clustering"
"Principal component analysis"	"Converts higher to lower dimensional data by possibly retaining significant information with orthogonal transforms""	"Dimensionality reduction"

compared to any single machine learning approach. Deep learning over performs in comparison to other conventional learning approaches in processing and quantifying complex data, such as images, text, and additional unstructured information.

Figure 13.8 shows the most popular supervised learning algorithm adoption rate in a medical context, and the most popular among them is S.V.M. and neural networks. Thus, when compared to unsupervised, supervised learning supports precise relevant clinical results, and hence A.I.-based practices for healthcare generally apply supervised learning.

13.5 DIGITAL TRANSFORMATION AND INTEROPERABILITY

Technologies such as 5G, cloud computing, artificial intelligence, natural language processing, and the Internet of Medical Things can help streamline and synchronize healthcare delivery with changing consumer needs. The increased use of Data-as-a-Platform (DaaaP) to extract information from patient records will be of concern to most healthcare organizations. If virtual healthcare increases flexibility and visibility, undoubtedly the organizations will need to keep investing in safety resources and facilities to detect and keep threats at bay.

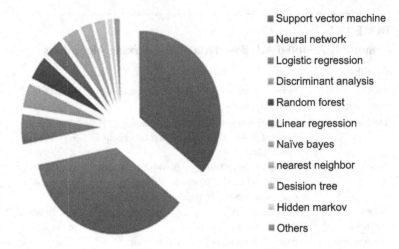

FIGURE 13.8   Adoption rate of various A.I.-based machine learning algorithms.

Five key measures to drive the digital transformation of healthcare:

i. **Build a comprehensive health I.T. system for data management, access to medical data, and knowledge exchange**

   Infrastructure should provide stable network access and ample data storage space; increase data management by shifting medical data to cloud providers; and deliver real-time remote access through device integration. Data should be easy to use for healthcare organizations, allowing them to pay just for what they need, such as software, apps, and technology resources.

ii. **Implement functional electronic medical records and invest in the fundamental infrastructure required for change**

   Organizations have to resolve fragmentation in the quantities, types, and capabilities of their E.H.R. systems to increase access to and use data from other vendors both externally and internally. Other critical technologies required to reach an appropriate degree of digitization include e-prescription, e-diagnostics, e-referral, e-recording, and e-discharge; safe information systems; digital imaging; point-of-care (P.O.C.) diagnostic tools; and electronic devices, patients, and personnel monitoring. They are built to maintain patients' health and improve clinical work processes by accelerating the flow and consistency of information and facilitating timely, responsive treatment.

iii. **Addressing the problem of interoperability**

   The interoperability of healthcare is incredibly complex and depends on the ability to create compatibility and safe data exchange between various and sometimes incompatible I.T. networks (and organizations). Healthcare professionals and their collaborators should consider partnering together to create an atmosphere to help lift the standard and enhance how they

exchange knowledge efficiently and effectively. Consistent use of data formats and transparent A.P.I.s and protocols will also help to increase con-
nectivity to and interoperability of E.H.R. systems by allowing data flow
from various applications and networks while upholding privacy and security requirements.

iv. **Build a robust governance system to promote a culture of digital
transformation**

Healthcare companies have to comply with the regulatory standards that
apply to information and analytics in general, and data protection and privacy in particular. Both healthcare systems and healthcare organizations
should provide a governing structure that allows for the implementation
and acceptance of secure, legal, and efficient data-driven healthcare technologies, and clarification on the entity's approach to information control,
patient consent, and patient care.

v. **Build technical communication skills and develop computer skills
among staff and patients**

The effective adoption of digital technologies needs leaders with a
strong view of the role of emerging technology in the advancement of care
delivery; healthcare personnel who are respected, motivated, and educated
in the use of technology; and patients who are motivated and assisted in
improving their digital literacy skills.

With various hurdles, major advances have been made in the digital transition
towards healthcare, which is expected to continue in 2020 and even beyond.

With technology adoption, healthcare services will see a transition in data processing from retaining databases to extracting information that can be monetized
and promoting developments in areas such as patient clinical care and value-based
care. During this development, digitalization in healthcare presented by existing traditional systems, the expense and sophistication of emerging technology, and ever-
changing customer demands and scenarios will continue to pose challenges, and
cyber security will remain a primary concern.

13.6   LIMITATIONS AND FUTURE PERSPECTIVES

To date, computers in medicine have demonstrated many critical applications for
both physicians and patients. However, it is a daunting challenge to control these
algorithms. A few assistive algorithms have been regularized by the U.S. Food and
Drug Administration (F.D.A.), but there are currently no requirements for uniform
approval. On account of that, people designing algorithms to be used in hospitals
aren't always physicians who treat patients.

In some situations, computer scientists may need to learn more about science and
practitioners would need to think about the roles for which a particular algorithm is
or isn't suitable.

Although A.I. can assist with diagnosis and simple clinical procedures, it's hard
to picture robotic brain surgery, for example, where doctors often have to change

their strategy when they diagnose a patient. In this respect, the ability of A.I. in medicine already exceeds the strengths of A.I. in medical care. Clarified F.D.A. recommendations, however, can help define algorithm specifications and might lead to an increase in clinically implemented algorithms.

In addition to limitations to F.D.A. acceptance, A.I. algorithms can also face difficulties in gaining patient interest and acceptance. Without a firm knowledge of how the algorithm works for clinical use, patients may not accept it to fulfil their medical needs. If required to choose, patients would rather be misdiagnosed by humans than an algorithm even where the algorithm outperforms clinicians. Appropriate decision-making is a function of the data system used as feedback, which is vitally essential for the proper application. With misleading data, algorithms can generate inaccurate results. It is also likely that individuals developing an algorithm will not realize that the data they furnish is deceptive until it is too late, and algorithms have induced medical malpractice.

Finally, the authors conclude that these obstacles are worth attempting to resolve to uniformly improve the precision and efficacy of therapeutic procedures for multiple diseases.

Future perspectives: Figure 13.9 represents the future perspectives of transitional medical research which includes financial operations and performance improvement; care model innovation; digital transformation and interoperability and the future of work.

Financial operations and performance improvemetn

Health systems are working to achieve financial sustainability by reducing the cost to delicer and finance high-quality and effective care for patients.

Care model innovation

New strategies, capabilities, and technologies are shifting health systems' focus from providing episodic, acute care to keeping people healthy.

Digital transformation and interoperability

Exponential advances and interoperability in digital technologies are helping clinicians deliver health care services in ways that consumers prefer to receive them.

Future of work

An aging population and shortage of skilled clinicians are changing the future of work in health care and increasing the importance of sourcing, hiring, training, and retaining skilled workers.

FIGURE 13.9   Future perspective of transitional medical research (original).

13.7 KEY-POINTS DRAWN

- Translating basic A.I. research into digital diagnostics depends on A.I. software and data sharing mechanisms.
- While maintaining personal data, ensuring that A.I. algorithms are robust and efficient in routine clinical practice, and ensuring consistency in integrating A.I. tools in conventional clinical practices.
- An A.I. environment in which radiologists, their technological societies, academics, designers, and government regulators cooperate, participate, and encourage A.I. in clinical practice would be crucial in transforming basic A.I. science into clinical research.
- A.I. tools will be a significant catalyst in the forthcoming clinical area where A.I. tools enable physicians and other diagnosticians to access enormous resources that can inform more precise diagnosis and classify patients at risk of substantial disease.

13.8 CONCLUSION

The emergence of the new technologies has witnessed the rise of data-ridden patient care services worldwide by improving the disease diagnosis quality, treatment, and management over various levels. With the next generation, healthcare systems and medical research will never be the same with the launch of A.I. support to imitate human cognitive activities and support a revamped paradigm of medical care with growing healthcare data and the rapid development of analytical techniques.

Throughout the study, we have discussed the critical applications of A.I./M.L. in translational medicine and further emphasized the significant limitations and bottlenecks of every application, and the prerequisites that must be addressed before the efficient application of these technologies in translation research or the clinical setup. Although the adoption of A.I./M.L. in medicine is accepted with high interest, overall additional care should be taken to anticipate incomplete rollout and the damage they could cause to patients due to flaws in these concepts.

We have explored the current state of healthcare A.I. applications and their future vision and discussed A.I. applied to different (structured and unstructured) variants of medical care information.

The most common A.I. approaches like machine learning models for structured knowledge and modern learning methods for unstructured knowledge have been adopted in the area of medical research. While A.I. technologies are capturing a significant hold in translational research, real-life implementation barriers still exist. The first challenge arises from the legislation because there are no criteria for the evaluation of the safety and efficacy of A.I. systems in the existing regulations. This is followed by the data-sharing challenge and the need to train (continuously) clinical trial data. However, after the implementation of historical data with initial training, the continuous data supply becomes a significant issue for further system development and enhancement.

Finally, there is the considerable temptation to impose a varying set of parallel computations on a substantial growing volume of data, but still there is an extreme need first to appraise the critical biological-related queries and the information to answer them instead of blindly adopting M.L. concepts for any dataset. However, M.L. developments will offer better analytical approaches to overcome the complexity found in biomedical research, making it possible for more accurate, data-driven decision-making to deliver the next century of therapeutic and diagnostic techniques to patients.

Further enhancements in methods like 3D bio-printing, gene editing, regeneration of cells/tissues, and the adoption of A.I. allow human genetics to be appreciated and enhance translational research to unprecedented levels.

"Thanks to A.I. and more can be achieved further by connecting A.I. (Artificial Intelligence) along with H.I. (Human Intelligence)."

BIBLIOGRAPHY

1. Briganti G & Le Moine O, "Artificial Intelligence in Medicine: Today and Tomorrow", 05 February 2020, https://doi.org/10.3389/fmed.2020.00027.
2. Orth M, Averina M, Chatzipanagiotou S, Faure G, Haushofer A, Kusec V, et al., "Opinion: redefining the role of the physician in laboratory medicine in the context of emerging technologies, personalized medicine and patient autonomy ('4P medicine')", *J Clin Pathol*. 72:191–7, 2019.
3. Esteva A, Robicquet A, Ramsundar B, Kuleshov V, DePristo M, Chou K, et al, "A guide to deep learning in healthcare". *Nat Med*. 25:24–9, 2019.
4. Topol EJ, "A decade of digital medicine innovation". *Sci Trans Med*. 11:7610. 2019, doi: 10.1126/scitranslmed.aaw7610.
5. Topol, EJ, "High-performance medicine: the convergence of human and Artificial Intelligence". *Nat. Med*. 25, 44–56, 2019.
6. Shah NR, Health care in 2030: "will Artificial Intelligence replace physicians?", *Ann Int Med*. 170:407–8. 2019, doi: 10.7326/M19-0344.
7. Verghese A, Shah NH, & Harrington RA, "What this computer need is a physician: humanism and Artificial Intelligence". *JAMA*. 319:19–20, 2018, doi: 10.1001/jama.2017.19198.
8. Price WN, Gerke S, & Cohen IG, "Potential liability for physicians using Artificial Intelligence". *JAMA*. 322:1765–6, 2019, DOI: 10.1001/jama.2019.15064.
9. Dorado-Díaz PI, Sampedro-Gómez J, Vicente-Palacios V, & Sánchez PL, "Applications of Artificial Intelligence in cardiology: The future is already here". 72:10, 65–75, 2019, DOI: 10.1016/j.rec.2019.05.014.
10. Niel O, & Bastard P. "Artificial Intelligence in nephrology: core concepts, clinical applications, and perspectives". *Am J Kidney Dis*. 74:803–10, 2019, DOI: 10.1053/j.ajkd.2019.05.020.
11. Vamathevan, J. et al, "Applications of Machine Learning in drug discovery and development". *Nat. Rev. Drug Disco*, 2019, https://doi.org/10.1038/s41573-019-0024-5.
12. Norgeot, B., Glicksberg, BS & Butte, AJ, "A calls for deep-learning healthcare". *Nat. Med*. 25, 14–15, 2019.
13. Yang YJ, & Bang CS, "Application of Artificial Intelligence in gastroenterology". *World J Gastroenterol*. 25:1666–83, 2019, DOI: 10.3748/wjg.v25.i14.1666.

14. Acampora G, Cook DJ, Rashidi P, & Vasilakos AV, "A survey on ambient intelligence in health care". *Proc IEEE Inst Elect Electron Eng.* 101:2470–94, 2019, DOI: 10.1109/JPROC.2013.2262913

15. Watson DS, Krutzinna J, Bruce IN, Griffiths CE, McInnes IB, Barnes MR, et al, "Clinical applications of Machine Learning algorithms: beyond the black box". *BMJ* 364:l886, 2019.

16. He J, Baxter SL, Xu J, Xu J, Zhou X, & Zhang K. "The practical implementation of Artificial Intelligence technologies in medicine". *Nat Med.* 25:30–6, 2019.

17. "Artificial Intelligence and Machine Learning in Software as a Medical Device". U.S. food and drug administration. https://www.fda.gov/medical-devices/software-medic aldevice-samd/artificial-intelligence - and-machine-learning-software-medicaldev ice#whatis; July 19, 2019.

18. Merk D, Grisoni F, Friedrich L, & Schneider G, "Tuning Artificial Intelligence on the de novo design of natural-product-inspired retinoid X receptor modulators". *Commun Chem.* 1. 2018, https://doi.org/10.1038/s42004-018-0068-1.

19. Gupta A, Müller AT, Huisman BJH, Fuchs JA, Schneider P, & Schneider G, "Generative recurrent networks for de novo drug design". *Mol Inform*; 37:1700111, 2018, https:// doi.org/10.1002/minf.201700111.

20. Popova M, Isayev O, & Tropsha A. "Deep reinforcement learning for de novo drug design", 2018 https://doi.org/10.1126/sciadv.aap7885.

Index

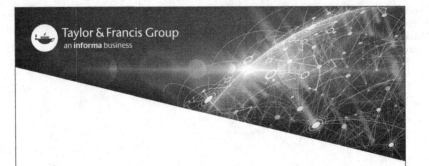